Work at the Boundaries of Science

Work at the Boundaries of Science

Information and the Interdisciplinary Research Process

by

Carole L. Palmer

University of Illinois at Urbana-Champaign

Springer-Science+Business Media, B.V.

A C.I.P. Catalogue record for this book is available from the Library of Congress.

ISBN 978-90-481-5884-3 ISBN 978-94-015-9843-9 (eBook)
DOI 10.1007/978-94-015-9843-9

Printed on acid-free paper

Originally published by Kluwer Academic Publishers in 2001.
Softcover reprint of the hardcover 1st edition 2001

CONTENTS

Preface	vii
Chapter 1: The Context of Interdisciplinary Science	1
Chapter 2: Information Accumulation	29
Chapter 3: Conditions of Boundary Crossing Research	61
Chapter 4: Accommodating Interdisciplinary Research	97
Chapter 5: Conclusion	123
Appendix	147
References	153
Name Index	165
Subject Index	169

PREFACE

> Knowledge about knowledge has a peculiar multiplier or leverage effect on the growth of knowledge itself. The more we know about learning and the transmission of knowledge, and the more we know about the processes by which knowledge advances at the frontiers, the more efficient will be the use of resources, both in education and in research.
>
> —Kenneth Boulding, *Beyond Economics*

The value of knowledge about knowledge, as expressed in Boulding's statement above, conveys the basic premise behind this book. Understanding how knowledge is produced and communicated is key to creating better conditions for knowledge development. Information, as a constitutive element of knowledge, plays an important role in the leveraging phenomenon identified by Boulding. To improve information resources for the pursuit of knowledge, we must first learn how information works at the knowledge frontiers. The research fronts of science supply a microcosm where we can see researchers working on scientific problems. The real-world research problems that scientists address rarely arise within orderly disciplinary categories, and neither do their solutions. Thus, the information needed to solve complex research problems is distributed across disciplines and takes many different forms, physically and intellectually.

Our stores of information are diverse, scattered, immense, and unwieldy. Information may be amassed in records and documents kept by organizations or stored in libraries, archives, and information systems. Some information exists only in personal or organizational memories. Tempered by the social and cultural facets of its creation and use, not all information is visible or usable. For example, some information circulates widely while some lies dormant, and esoteric languages may make specialized information incomprehensible to those outside a discipline. Moreover, the organization of any collection of information reveals certain features while disguising others.

It seems a wonder that scientists manage to be productive and innovative amidst this complicated resource base. But, they do, and in turn their ways of working with existing information to create new knowledge tell us much about the enterprise of science and how to foster its advancement.

The empirical work that fed into this book began in 1992 when I got involved in a three-year federated research project investigating interdisciplinary scientific communication and research processes. The team members on that project were based in a number of social science fields—anthropology, communication, sociology, and my field of library and information science. United in our interest in the knowledge structures of scientific communities, we investigated a series of questions related to social and organizational aspects of communication, information use and exchange, and legitimacy and competition in interdisciplinary science. Due to our diverse backgrounds, we brought a broad perspective to the project and used a range of methods, including citation analysis, network analysis, survey questionnaires, and structured interviews. Pertinent results from that federated research, referred to as the "general survey" within the text, have been integrated into the analysis presented here. That project in its entirety formed a solid foundation for the independent research I undertook in subsequent years. As a sole investigator, I followed a similar line of questioning. My aim was to identify the critical components of the interdisciplinary research process to inform the development of advanced information systems and services for innovative problem solving and inquiry.

The applications I envisioned for this research are important, but information science is but one of a number of audiences concerned with the interdisciplinary research process. Research administrators and practicing scientists and scholars from a range of fields are interested in creating and working in environments that help researchers traverse intellectual, cultural, and organizational boundaries. Within the literature, this research fits in the growing field of knowledge studies that cuts across the social sciences and the humanities, and it contributes to the body of social and practice-based studies of science and knowledge production. It follows in the tradition of studies of science communication that have infused information science at least since Derek de Solla Price's *Science since Babylon* (1961) brought attention to the role of recorded knowledge in science.

This book is concerned with the relationship between interdisciplinary research and information. It is based largely on interviews with scientists about how they work across disciplines to solve research problems. While the disciplinary boundaries of science are progressively shifting and dissolving, they continue to bring needed structure to the world of research by differentiating consolidations of knowledge and expertise. But, depending on the problem at hand, any segmentation scheme can obstruct communication and information transfer and deter the mobilization of knowledge. In the practice of interdisciplinary research, disciplinary boundaries, like most other components of the research process, require a certain amount of management.

My investigation of the interdisciplinary research process focused on the strategies scientists use to manage boundary crossing information work. Studies of the practice of interdisciplinary work, as Weingart and Stehr (2000) suggest, yield more vital insights than studies that attempt to define disciplines or weigh the merits of interdisciplinary efforts. For this reason, I have not strictly delineated disciplines or tried to determine what mix of disciplines constitutes authentic interdisciplinary research. In fact, my analysis does not support any one conceptualization of interdisciplinarity, nor does it directly challenge any of the multiple ways that discipline crossing research has been portrayed in the past. Instead, I provide a close look at the practices and conditions that generate interdisciplinary science, corroborating that there are many ways that people and information move across boundaries and interact effectively during the course of complex and integrative scientific work.

It is important to note, however, that because the notion of interdisciplinary research has not solidified, debate about what it really means goes on. As Klein (1996) points out, "interdisciplinary" was included in the 1977 *Dictionary of Diseased English*, a compilation of words that had been "used with so serious a lack of precision" that they ceased to be effective for communication and serve only to "confuse or mislead" (Hudson 1977, xix). Over time, the term has become more commonplace but no less ambiguous. Other words used to describe the complex intellectual world also leave much to interpretation. We have an idea of the function of disciplines and subdisciplines—they bring order to researchers, students, methods, journals, and the like. And, while they commonly take form in academic departments and curricula, professional organizations, textbooks, and systems for classifying knowledge, it is nearly impossible to discern what exactly falls inside or outside a discipline at any given point in time.

To investigate the dynamics of interdisciplinary work, I turned my attention away from the bounds of the disciplines themselves and concentrated on the scientists and their interactions with people and materials outside their core research area. My approach makes conceptual distinctions between levels of disciplinary interaction within a loose hierarchy of synthesis. Within this frame of reference, a discipline represents the subject areas, tools, procedures, concepts, and theories of a stable epistemic community (Klein 1990). "Multidisciplinary" describes an additive juxtaposition of disciplines. For work to be considered interdisciplinary, it must bring together and synthesize material from more than one discipline. Of course, some amount of synthesis will generally take place as an interdisciplinary project is carried out, but the heterogeneity of the subject areas involved and the degree of integration can be quite variable. Transdisciplinary, a concept that is currently receiving considerable attention, implies a higher-level synthesis. Transdisciplinarity is a means, or a way of working and acting, that produces holistic, integrative knowledge. This genre of research is problem-centered, participatory, and involves multiple stakeholders (Klein 2000, forthcoming.)

Throughout the text, I consistently apply the terminology favored by the scientists during the interview sessions. As a result, the term interdisciplinary pervades throughout the book. It clearly still resonates with scientists, and it continues to be used widely in academe to describe cross-cutting programs and initiatives. I have used the term cross-disciplinary when the process of integration is not meant to be implied to a particular situation. Since the term transdisciplinary was not part of the participants' vocabulary, it does not appear in descriptions derived directly from the empirical data.

To faithfully represent the interdisciplinary research process as conveyed by the scientists, I interweave their words with my analysis to tell the story of how interdisciplinary work happens in their research world. Chapter 1 lays out the context of interdisciplinary research locally at the research site, and conceptually as conveyed in the literature. Here I also sketch out a model of interdisciplinary research modes, a primary outcome of the study that provides structure for talking about the details of work practice. In Chapters 2 and 3, I build a narrative account of boundary crossing information work and the interdisciplinary research environment as experienced by the researchers. These data-rich chapters cover the practices researchers use to gather and use information, and to learn, build, and disseminate new knowledge. Here we also come to understand the scientists' orientations to their fields of study and the institutions and cultures in which they work. Chapter 4 covers the larger processes of research—how it happens and what needs to be in place for progress to be made in terms of both personal and organizational resources and strategies. In the concluding chapter, I discuss the implications of the multiple modes of interdisciplinary work and priorities for developing supportive environments and information systems for interdisciplinary research.

Earlier reports of this research and previous versions of some of the material in this book have appeared in the following publications: "Information Work at the Boundaries of Science: Linking Information Services to Research Practices," *Library Trends*, v. 45, no. 2 (1996): 165-91; "Ways of Working and Knowing across Boundaries: Research Practices of Interdisciplinary Scientists," in *Finding Common Ground: Creating the Library of the Future without Diminishing the Library of the Past*, edited by Cheryl LaGuardia and Barbara A. Mitchell, New York: Neal-Schuman, 1998; "Structures and Strategies of Interdisciplinary Science," *Journal of the American Society for Information Science*, v. 50, no. 3 (1999): 242-53; "The

Information Connection in Scholarly Synthesis," in *Discourse Synthesis: Studies in Historical and Contemporary Social Epistemology*, edited by Raymond G. McInnis. Westport, CT: Praeger, 2001.

I regard many people as significant contributors to this book. Most importantly, I am indebted to the scientists who participated in this research. They graciously volunteered their time, but their introspection and openness about their experiences are what made this work possible. Geoffrey Bowker's guidance and feedback carried me through the many stages of development and early drafts of the manuscript. His brilliant insights made their mark on the text, while his brilliance as a mentor endures in my researcher psyche. At several stages during the course of this project, Julie Thompson Klein provided invaluable assistance by sharing commentary, texts, and her vast knowledge in response to my queries. I am grateful to the anonymous reviewers and local readers for their many helpful comments on early versions of the manuscript. I wish to thank Linda Smith and Pauline Cochrane for their comments on the text and for urging me to pursue publication, Leigh Star for her advice on later stages in the publication process, and especially Leigh Estabrook for her sage advisement and consistent confidence in my particular blend of information science. In the final stages of production, Neil Robinson showed great devotion and evenhandedness in his attention to the details of manuscript preparation.

The years of work that went into this book have been sustained by my wonderful and generous colleagues at the University of Illinois who have shown unfaltering support and kept me intellectually engaged and in good cheer. Finally, I dedicate this book to Ed, Evan, Laurel, and Nina, who regenerate all my academic pursuits. They are my most genuine advocates and my most glorious exploit.

Chapter 1.

THE CONTEXT OF INTERDISCIPLINARY SCIENCE

If we could visually trace how the entire stock of knowledge has expanded over the past half century, the spread of the exterior boundaries and the increase in overall mass would be considerable. An equally striking change would be seen in the internal geography. Knowledge has been in a state of flux—a continual process of reconfiguration, with existing subject domains merging and seceding, and new ones emerging. As the structure of knowledge grows in both scope and specificity, the conduct of research is also changing. Increasingly researchers are importing and exporting information, techniques, and tools across disciplinary boundaries and working together to apply more powerful and sophisticated approaches to the questions they ask. They manage to continue to solve important research problems by adapting their methods of inquiry to the breadth and complexity of knowledge.

The importance of interdisciplinary research has been recognized since the Social Science Research Council was established in the mid-1920s (Klein 1996, Fisher 1990). In recent decades scholars from a number of fields have begun to take a close look at where, how, and why it is done, and to what end.[1] Path breaking ideas are said to emerge through the cross-referencing of ideas across disciplines (Turner 1991), and it has been proposed that disciplinary boundaries are the fault lines that conceal future scientific revolutions (Fuller 1988). It has even been argued that we have passed into a second mode of knowledge production, a transdisciplinary phase that not only transcends traditional disciplinary frameworks but gives rise to more socially responsible creators of knowledge (Gibbons et al. 1994). There are people who take a heterogeneous and non hierarchical approach to research, and there are places that strive to support this ideal. This book is about one such place. Through the study of scientists at an interdisciplinary research site, we will see that an organization constructed to promote interdisciplinary exchange and synthesis may, for better or worse, still be bound by perspectives and structures associated with traditional disciplines.

Interdisciplinarity is not simply about how research and education are arranged within institutions. It is not an isolated or esoteric phenomena (Jantsch 1980) but a pervasive approach to inquiry found within disciplines and outside them. "Disciplines now routinely experience the push of prolific fields and the pull of strong new concepts and paradigms" (Klein 1996, p. 56). Interdisciplinary work has become an essential part of the ongoing process of knowledge production, "from the point of making claims to legitimating practices and judging outcomes," to forming new hybrid disciplines (57). Yet little is known about the process itself, about how information and knowledge are transferred between communities and mobilized to address research problems. How is information gathered, combined, and disseminated across intellectual boundaries? How do researchers overcome or work around the physical, social, and cultural barriers that exist between fields of research? As an information scientist, my work is aimed at understanding the processes by which people become informed, expressly for application to the design of information systems and services for particular communities of users. But, in effect, how information and knowledge are used, organized, and produced is of interest to numerous fields of study, including communication, sociology, management, and education. It is also of practical concern to individuals involved in interdisciplinary research and education, as well as universities and governmental agencies where decisions about funding and development of infrastructure for research are often made without a thorough understanding of what is necessary for research work to be performed effectively.

As new information technologies are adopted in our research institutions, it is easy to see how they can reduce some constraints, especially those that have to do with time and the location of work. There is no doubt that computers are essential tools for communication, the display and manipulation of information, the analysis of data, and the simulation of systems. Nevertheless, "integration is a human action" (Klein 1996, 218). There is no overlay of technology that can permeate the organizational, intellectual, and linguistic barriers that constrain the flow and exchange of information across disciplines. Information integration is knowledge work done by people. This book is about how a group of academic scientists manages to work across dynamic, overlapping, and mutable knowledge domains. It is about the skills, strategies, and tools they use to bring together the resources they need to work on scientific problems. Their practices and experiences reveal what is required for the conduct of interdisciplinary research, ultimately leading us to see why emergent intellectual communities, and their communication and learning patterns, are just as important as machines and money.

The dual functions of research and teaching complicate the academic research environment, creating a tension that causes both fragmentation and integration (Clark 1995). As research becomes more interdisciplinary, current academic subject frameworks become increasingly ill suited to how research is really done. The disciplines may still be adequate for coordinating teaching activities within a university, but they are misleading simplifications of research areas and the intellectual domains that sustain them (Geertz 1983, Becher 1990, Pinch 1990). Interdisciplinary research centers are one of the various types of physical and administrative frameworks constructed within universities to realign the skew between academics and research. The site for this study, hereafter referred to as "the Center," is an institute of this kind. It is a well-supported unit at a large public university with programs spanning the physical sciences, engineering, computational science, the life sciences, and the behavioral sciences. In 1995, when the bulk of the data was collected for this study, nearly two dozen academic departments, ranging from physics to anthropology, were represented in the faculty membership. At the Center, then and now, interdisciplinarity is not only accepted, it is expected.

The Center was founded "on the premise that reducing the barriers between traditional scientific and technological disciplines can yield research advances that more conventional approaches cannot."[2] This tenet was the basis for both the architectural design of the building and the assemblage of research groups. The facility, funded by an alumnus donation and state appropriations, is the second largest, and, in many ways, the most impressive building on the campus. There are two main wings: one with more than 200 offices, and a second laboratory wing with state-of-the-art computing, simulation, visualization, laser, and magnetic resonance technology. The two sections are connected by a series of bridges above a public atrium. There are public spaces for more casual group interaction, and ample meeting rooms for formal gatherings and conferences. The physical layout and the programs encourage open exchange of information, as do the many national and international workshops and conferences hosted by the Center. The facility was engineered to offer what one administrator called "bold communication possibilities."

When the study was conducted, the Center's research force was made up of approximately 150 faculty members; dozens of visitors from other universities, government, and industry; countless postdoctoral fellows, graduate and undergraduate students; as well as a full line of administrative, computing systems, and support staff. In addition to the group of researchers I interviewed for the project, I was able to enlist other participants, including an administrator and an artist in residence. Throughout the text I refer to them as the Organizer and the Humanist. The Organizer contributed detailed information about the Center's intended goals and early development, and the Humanist gave valuable interpretations of the research environment from the perspective of an informed outsider. Most of the analysis in the chapters that follow is presented from the researchers' point of view. Here I give a preliminary introduction to the Center from the unique perspectives provided by the Organizer and the Humanist.

The Organizer

> I've always thought of this thing as a big experiment. Just put something there at time equals zero, turn it on and let's see what

> happens. Some people fall out and some people come in . . . let it run a while and watch to see what happens.

The Organizer was an instrumental figure in the conception and realization of the Center. He played a key role in developing the initial proposal and in planning the building. After the Center opened, he was involved in its administration for more than five years. The primary goal of the Center was "first class research as judged by the people that work in the field and by the peer review process." His original vision was of a place unrestricted by departmental barriers, "a big, happy playground," where people could mix and learn new things. In retrospect, it didn't turn out exactly that way. For instance, he had expected that the cafeteria would be full of people having lunch and telling one another about their research, but on most days the lunch crowd comes from outside the building, and the Center members who are there tend to sit in the same groups. He had also been excited about the cross-disciplinary seminar program that was implemented, but attendance proved to be consistently poor. He discovered that "people don't come in droves to hear very general, broadly interesting talks." They are just too busy. Although all his hopes for the Center did not come to pass, he remains convinced that it *does* work, just "not at the high powered, ideal level that you might think."

The initial research programs were drawn from a pool of proposals submitted by "self-organizing groups" on the campus. Those selected fit with the original mission of the Center that was aimed at linking the efforts of researchers in many disciplines who were engaged in the quest for a better understanding of complex systems. The explicit mission changed over the years. Later promotional materials were less specific about the target areas of research, stating that the Center's goal is "to foster interdisciplinary work of the highest quality in an environment that transcends many of the limitations inherent in traditional university organizations and structures." This is a better representation of how the initiative unfolded. The planners were seeking broadly interdisciplinary groups that "spanned more than one area" and included faculty from multiple departments. Groups also had to demonstrate that the Center would add value to what they hoped to accomplish. The Organizer admitted,

> We were not completely successful. Many of the programs we chose were not really all that interdisciplinary but had the potential for reaching out and taking advantage of this opportunity to interact with these other groups.

Some were multidisciplinary in their composition; others were subunits of an academic department.

The Organizer expected that bringing together a collection of interdisciplinary research programs would produce longer range intellectual connections between them. The Center would provide the structure, encouragement, and status necessary to move the existing groups to another level of interaction, but it was not intended to be a permanent home for any programs or faculty. The membership was to change depending on who could benefit most from the facility, and, in turn, who could contribute the most to the mission of the Center. Not surprisingly, it was difficult to manage this rolling membership system.

> It's not something that spontaneously works. . . . All the manager can do is try to create an environment that is sort of proactive in stimulating these things, providing such inducements as can be provided through allocations of resources, but not with a heavy hand.[3]

Subsequent administrators have worked to make the overall research agenda more focused by defining general areas of research activity, a move that provides more thematic structure while maintaining the initial emphasis on cross-group contact and long range connections between disciplines and specializations.

The Humanist

> There is still that element . . . where you say, well what *is* the common denominator? How *do* you get these privileges and these resources? Who gets to play and who doesn't?

The Humanist was the first person appointed to the Center who did not have some kind of "quasi-mechanical" relationship with the facilities. Known to be a "cross-boundary freak," he was not out of place at the Center but was pondering the same question as many working in the neurosciences and artificial intelligence—what is the shape of knowledge? The artist in residence position was served up to him "as a free sample," and as a regular member of the institution he was able to take full advantage of the available resources. He was intrigued by the prospect of working in a place that promoted interdisciplinarity. The environment provided both stimulation and seclusion—qualities also appreciated by the scientists. The opportunity to do creative work in an isolated "playground" was most inviting. It is interesting that both the Humanist and the Organizer used the playground analogy to describe the Center. I do not know where the metaphor originated, but according to the Humanist, it never surfaced in conversations between him and the Organizer.

The work the Humanist completed while in residence was greatly influenced by the Center's atmosphere, rich with impressions of the place, the people, and their activities. At night when it was quiet, he liked to roam the halls to experience the sounds and "study the posters about nerve cells and microchips." Then he would go to his office to work at the computer. At official social functions he was an outsider, the token humanist. His struggles to converse with his Russian officemate and other encounters with scientists felt like "exercises in coherence," but they were not unlike the communication difficulties he observed between different scientific groups within the Center. Over time, and with the help of some of the scientists, he learned the "local colloquial" necessary for coherent exchange of ideas in the "community." As we will see, this scenario is very similar to the researchers' experiences with disciplinary vocabularies.

At the beginning of his time at the Center, the Humanist thought of it as a self-contained "city within a city." It had everything to attend to both physical and intellectual needs, including food, a library, and "any number of people with beds in their labs or offices." The Center seemed to house a separate population, a group not completely assimilated into the rest of society. Approaching the building from different directions on his way to the office, he was struck by its strong barrier quality and its "territorial, archaic feel." It gave him the sense that people behaved differently beyond those walls. Comparing it to other parts of the campus, he recognized the "inequitable privilege" of having an office there.

The Center, with all its new possibilities, brought on an instant change in the Humanist's sensibility; it had the ability to "uncramp." This feeling came partly from the material advantages of being a member, but it was more than that; it was exciting in unexpected ways. Just looking at all the new fields of study listed on the main entrance directory was stimulating. After having moved back to the part of campus populated by the humanists and social scientists, he observed: "these [new fields] are the kinds of things they try to buffer us from on this side of campus." He was sensitive to the indirect influences within the Center, those things that "buzz in inaudible frequencies," like the directories or the titles on the lecture announcements that get put in the mailboxes each week. "Why can't people in the humanities departments get those fliers and be thinking about those titles?" Another distinct and impressive change for the Humanist was the Center's flourishing scientific apprentice system that seemed to allow graduate students to guide the direction of research. He noted that students' roles were quite different on the "other side of campus."

Over the course of our first discussion, the Humanist referred to the Center as a city, an orchestra, a living organism, cognition, and the chef's special. Reflecting on these descriptions in our second meeting, he settled on the organism as the best analogy because it acknowledged the importance of the lives of the individual cells.

> The orchestra metaphor would probably be more accurate if we restricted it to the three or four minutes of tuning up. Everybody is making noise, but nobody has come out to tap on the podium yet. The chef's special is for when you want to be cynical . . . an excuse to throw more money at people in an ad hoc way.

He also favored his original "city within a city" image because it exemplified the level of activity and the mix of "all kinds" of people. On the individual day-to-day level, it appeared to him that what the scientists do may not be that different from what they have always done. The overall organism, however, represented something quite significant—progress toward "hierarchical convergence" in science, an attempt "to reintegrate the reductionist program" through a "convergence of model making."

CONCEPTUALIZING INTERDISCIPLINARY INTERACTION

The interdisciplinary reconfiguration of knowledge has been portrayed in many ways: conjuring images of overlapping, blurring, displaced, and shifting boundaries; merging, fusing, and intersecting domains; and mingling and migrating individuals (Campbell 1969, Mulkay 1974, Chubin 1983, Hoch 1987, Klein 1993). None of these descriptions completely explains all the levels of activity and the interplay between people and scientific communities at the Center. A comprehensive account would document the expanding knowledge base of individuals, the creation and distribution of knowledge within institutions, the embodiment of knowledge within intellectual communities, the transfer of knowledge between these entities, as well as the vital role of information throughout these processes. We can begin to formulate a general framework for understanding interdisciplinary research by bringing together some of the key concepts that have been used to represent how knowledge domains interact and coalesce. The concept of *knowledge units*, applied in relation to the *core*, *scatter*, and *hybridization* of knowledge, effectively illustrates the social and intellectual exchanges and reformations in the general landscape of knowledge. The reconfiguration of knowledge can also be understood at the level of research practice, as a cycle of knowledge *accumulation* that involves *boundary work* and *boundary objects*, research activities and materials that expedite the process of knowledge accumulation.

Fisher's (1990) notion of knowledge units is useful for thinking about the intellectual interaction involved in cross-disciplinary inquiry. Building on work from the sociology of knowledge and sociology of education, Fisher conceives of integrated knowledge as weakly classified units with flexible boundaries that allow open exchange between people.[4] The scope of a knowledge unit is variable; it can be "a discipline, a subdiscipline, or a group of disciplines" (98). For example, general classifications such as mathematics and engineering are typical disciplinary knowledge units, whereas optics is a smaller unit within physics. Integrated knowledge may be an agglomeration of disciplinary, specialized, and fragmented units, but the various units must exist in open relation to each other. Extending this idea to interdisciplinary knowledge, an integrated cross-disciplinary field will have a core of knowledge that is made up of multiple, interdependent units. The units may be disciplines, subdisciplines, or less developed subject areas, and some or all of these units may be equal in emphasis. Interdisciplinary fields evolve and integrate over time as researchers work with, exchange, and synthesize knowledge from multiple units.

Studies of scientific communication suggest that knowledge units interact and influence each other through scatter (Chubin 1976). Drawing on Crane (1969) and Bradford (1953), Chubin asserts that without the scatter of knowledge, scientists would be isolated in small groups that only speak to, read, and cite each other. Even scientists who work from a highly disciplinary core draw from the periphery of their domains, deriving "innovations from the margins" (448). Knowledge development within the core promotes growth and cumulation, and scatter outside the core keeps science from being "a sect-like phenomenon" (Crane 1969, 349). While fragmentation and scatter can inhibit the general "utilization" of knowledge (Beam 1983), they are essential for the discovery and innovation that stimulate the growth and evolution of knowledge.

Dogan and Pahre (1990) propose that knowledge is reconfigured through a "specialization-fragmentation-hybridization" process whereby knowledge units merge and develop into various informal and institutionalized structures. Informal hybrids can take the form of cross-disciplinary research topics or networks of interpersonal contacts. More permanent institutionalized hybrids are realized in cross-departmental academic programs, emergent fields of study, and interdisciplinary research organizations. The field of psychology can be thought of as a conglomeration of hybrids that includes social psychology, physiological psychology, political psychology, behavioral pharmacology, and cognitive science. In particular, cognitive science, which draws from linguistics, computer science, neuroscience, and philosophy, illustrates psychology's interconnections with outside fields (Dogan and Pahre 1990). Biophysics is another obvious case. Physics tools have been involved in the investigation of biological problems since the development of the optical microscope. Today, theoretical physics is fundamental to biological studies of proteins, genetic materials, and the multicellular organization of the brain

(National Research Council 1986). Cognitive science and biophysics are among the more highly developed scientific hybrids, as evidenced by their strong presence at the Center and within the schema of conventional academic departments. They have been firmly classified in structures of knowledge and research. Many newer, evolving interdisciplinary areas, such as environmental studies and urban studies (Klein 1996), do not yet have a stable place in categories of knowledge or institutions where knowledge is produced and disseminated.

The core and scatter dynamic represents the intellectual distribution and circulation of knowledge, while hybridization depicts the conflation of knowledge as manifested in institutions and organizing structures. These concepts are useful for tracking and interpreting the shifts and changes that continually affect disciplinary relations. Cores of knowledge have established boundaries in organizations and systems of categorization; scatter moves knowledge beyond the perimeter of those boundaries; and hybridization reconstructs the boundaries. More palpable than these abstract notions, however, are the hard work and the obstacles that are a substantial part of interdisciplinary research. The term scatter, in particular, suggests something much more random than the very deliberate and directed activities performed by the scientists at the Center as they transfer ideas, information, and themselves across boundaries. In their day-to-day work the scientists set priorities and develop routines that help them to move beyond their core into other domains to gather and distribute the wares of research.

The research processes that build interdisciplinary knowledge involve core development, scatter, and hybridization. As Latour (1987) demonstrates, a working definition of knowledge must be based on an understanding of what it means to gain knowledge.

> Knowledge is not something that could be described by itself or by opposition to ignorance or to belief, but only by considering a whole cycle of accumulation: how to bring things back to a place for someone to see it for the first time so that others might be sent again to bring other things back. (220)

Latour uses the example of expeditions by early French navigators to illustrate the nature of accumulation. The navigators gained what they needed to know for their travels through "kings, offices, sailors, timber, lateen rigs, spice trades, and a whole bunch of other things not usually included in 'knowledge.'" The navigators' cycle of accumulation included all the actions and interactions, exchanges and negotiations, objects and people that made it possible for them to carry out their work and document it in enough detail for someone else to follow their route at a later time. The next expedition would be charted—the water, the land, and the natives would no longer be unknown.

Knowledge accumulation in the practice of research is similarly dependent on a range of actions and interactions, exchanges and negotiations, objects and people, many of which are taken for granted or not easily observed.

Latour emphasizes the things within the knowledge cycle that enable information to be moved and combined, because they accelerate the accumulation process. Map making, printing presses, engraving of plates for scientific texts, projection systems, classification schemes in libraries, and, of course, computers work in this capacity (228). This relationship between the mobilization of information and the accumulation of knowledge is particularly salient in the context of contemporary research where networked computing environments allow information to be readily transferred and transformed. Information technologies are changing how information is moved and used, and for interdisciplinary scientists who need to accumulate knowledge from multiple disciplinary worlds, they offer new avenues for gathering divergent information and working across boundaries.

Gieryn's (1983, 1995, 1999) extensively developed concept of boundary work relates to the scientific practices that construct the social boundaries that separate science from everything else. Boundary work, in this sense, "is a strategic practical action" performed to monopolize, expand, and protect science (Gieryn 1999, 23). Fisher (1990) applied the idea to scientific activities more broadly, defining boundary work as "acts and processes that create, maintain, and break down boundaries between knowledge units," acknowledging, as does Gieryn, that it simultaneously "involves institutions and social structures" (98). Klein (1996) also extends the idea of boundary work beyond the role of demarcation to cover how boundaries are permeated and dismantled. In the context of interdisciplinarity, Klein defines boundary work as the

> . . . composite set of claims, activities, and institutional structures that define and protect knowledge practices. People work directly and through institutions to create, maintain, break down, and reformulate boundaries between knowledge units. (p. 1)

Their activities "attribute selected characteristics to particular branches of knowledge on the basis of differing methods, values, stocks of knowledge, and styles of organization" (Klein 1993, 185-186). Adopting Fisher's emphasis on knowledge units and their malleability, this variation on the concept of boundary work accounts for the professional authority and expertise central to Gieryn's conception, while capturing the crossing and reconstruction of boundaries that takes place as researchers break out of established domains to create new knowledge. My analysis is largely focused on the information and communication functions required for crossing boundaries in the accumulation process and on the specific resources and activities in the research environment that support this work. While these concerns are best understood within the context of how boundaries are broken down, the scientists at the Center also take part in building boundaries, for the two kinds of boundary work are inseparable in the real world of research.

Some of the things that assist in boundary work across disciplines can be considered boundary objects. As defined by Star and Griesemer (1989), boundary objects "inhabit several intersecting social worlds" and are able to "satisfy the informational requirements of each of them" (393). The concept is fitting in that traditional scientific disciplines, with their cohesive activities and subject concerns, function as social worlds (Gerson 1983), and certain objects in the research environment help people from these different worlds to come together to conduct research and make progress on scientific problems. Indeed, the researchers at the Center identified information mechanisms that embody this combination of stability and adaptability, such as scientists who are talented communicators, synthetic cross-disciplinary review articles, or a particular analogy that can crystallize an idea from an outside subject area. At the same time, other objects also traject boundaries but do not meet the strict criteria of a boundary object. These quasi-boundary objects exist primarily in one disciplinary world or consist of information that needs to be substantially adapted before it can be used in another world. While, the boundary object concept stresses cooperative tasks that allow separate social worlds to achieve goals in spite of boundaries (Gieryn 1995), not all boundary activities are overtly or actively cooperative. They may be competitive in nature, or, as we will see, some interdisciplinary scientists perform vital boundary crossing science independently.

Many elements that contribute to the interdisciplinary accumulation process are facets of or variations on the established boundary work and boundary object concepts. Here we are concerned with all the activities and resources that researchers use to traject boundaries—to rise above or get past barriers and move on with the work of accumulating information and knowledge across domains. The work activities and objects that assist in this process rarely stand as singular means for boundary crossing research. Instead, multiple components are employed in concert and over time to bridge boundaries and open channels. Some of the more important trajection elements are people, data, methods, and words.

People are involved in every aspect of cross-disciplinary work. "The big guys" loom over disciplinary territories long after they die, influencing the direction of science through their followers.[5] Colleagues give researchers a sense of place or belonging, and personal contacts are one of the most important vehicles for exchanging information across borders. Students play a versatile boundary role by functioning as human conduits for the passage of information. They are traded between labs, used as translators between theory and experiments, and sent as emissaries to other parts of academia and out into industry.

Frequently, concrete things are vital features of discipline crossing. Data (numbers) and data sources (rabbits) are shared between labs and sometimes brought together for comparative analysis. Banks of raw data are amassed and then added to by allied researchers. One research group may analyze molecules built by another distant group, with both sides bringing insights to the final results. It is common for apparatus to be borrowed and applied in new ways and to different types of data. New computational technologies are often combined with established disciplinary science to push the frontier of studies in an area, and computers are used to manipulate the boundary between the theoretical and the experimental. For many at the Center, data and theory are brought to life through computer modeling. The computer's role between disciplinary boundaries is less clear. At the very least, sophisticated computation may enable boundary crossing by producing models that can be applied broadly across sciences.

Methods move across boundaries in a number of ways. Researchers may bring techniques and procedures from different disciplines to their research problem. For example, one psychologist listed the following measures as part of his investigative repertoire: reaction time and accuracy measures from cognitive psychology, event related brain potentials from neuroscience, and magnetic resonance imaging from physics and chemistry. Several cases illustrated how experts trained in a methodological approach in one discipline use their expertise as an entry point into another disciplinary domain. A computational neuroscientist learned computer modeling and simulation as a physicist. He later transferred these skills to neurobiology, where he then contributed experimental and theoretical work by building on previous electronics experience and physical science training.

Words can also be the meeting point for different sciences. Metaphors can act as models, creating new frameworks for addressing scientific phenomena. A number of researchers talked about using metaphors as a tool to help groups of people from disparate backgrounds think about a problem in the same way. Words and concepts cross borders, and over time the vocabularies of different communities change and merge. As certain terms become more broadly applicable, there is more cross communication between disciplines. For example, a biologist working on charge separation may get the attention of a physicist interested in electron transfer or a biochemist working in catalysis who is interested in proton transfer. The flow of words seems to take place more readily through reading than writing. Many of the scientists read across disciplinary boundaries, while few make large leaps in their writing and publishing. Words are, perhaps, the most tenuous of boundary elements. They can generate cross-disciplinary understanding, but at the same time they can be a serious impediment to communication between scientific cultures.

Through this introductory inventory of trajection elements involved in research done at the Center, we can begin to see the range of activities and resources used by researchers to accumulate the scattered knowledge they need to work on their research problems. Although resources are often deficient as boundary objects and many activities make only small boundary work advances, synergistically these elements are as instrumental to interdisciplinary research as the laboratory equipment and the workforce of research assistants. Later chapters provide a detailed examination of boundary crossing research as practiced at the Center. By identifying what moves across boundaries during the course of research, and tracing how these things are gathered, translated, and transmitted by researchers, we will see how interdisciplinary work is conditioned by the information environment and its ability to support the accumulation of knowledge across disciplines.

STUDYING BOUNDARY CROSSING RESEARCH

Aspects of the interdisciplinary research process have been studied using a number of approaches. Based on a content analysis of the existing literature, Chubin et al. (1986) identified three major areas of study: those focusing on barriers to performance, examinations of cultural and organizational contexts, and works concerned with applications of interdisciplinary research. This study does not fall into any one of these categories, although the analysis does address the first two in some detail. Rather, it is an investigation of how scientists manage to synthesize the research of two or more disciplines—an area of focus suggested by Fuller (1988) for learning about disciplinary boundaries. The approach is similar to that used by Becher (1989) in his work on forms of knowledge communities in academe. While not interested in interdisciplinarity, per se, Becher also worked from a sociological perspective, drew on studies from information science, and used interview data to explore the nature of academic disciplines, their boundaries, specializations, and communication patterns.

The Center is an institutional niche that gives us a unique perspective on the structural and cultural contexts and confines of disciplines in academe. Knowledge work and research work are difficult to study, in part because they are less visible than other types of work (Star and Strauss 1998). To understand the creation of scientific knowledge, investigators must apply methods that go beyond the analysis of the end products that scientists create to follow the ongoing processes of research work (Latour 1987). Labs, institutes, and departments provide a place for examining practices, techniques, and instrumentation at the local level (Ravetz 1971, Latour and Woolgar 1979, Knorr-Cetina 1981).[6] Microapproaches of this kind fall into two categories: studies of students learning their discipline and those of scholars and scientists practicing theirs (Messer-Davidow, Shumway, and Sylvan 1993). This study falls into the second category, but it examines the orientations and work practices of scientists who, rather than being embedded within a discipline, have been awarded a special position outside the standard structures of academic research. Both quantitative and qualitative methods were employed to address three primary research questions:

What resources and activities are involved in boundary crossing research?
How do researchers accumulate knowledge (in the Latourian sense) outside their core specialization?
How does the research environment influence the accumulation process?

Two phases of data collection and analysis were conducted to examine interdisciplinary research within the framework of the above questions. First, it was necessary to identify individuals who could be considered active boundary crossers. I had previously been involved in a large-scale survey study of interdisciplinary research at the Center, and while collecting data for that project it became evident that there was considerable variance in the degree to which scientists crossed intellectual domains in their current work. As discussed earlier, *groups* of faculty were appointed to the Center based on their fit with the organization's mission. Some groups were more homogeneous than others, and in some cases, group members came from only one or two cognate academic departments. Moreover, some individuals clearly had more interactions with subjects and communities outside their core specialty. Using a theoretical sampling approach (Glaser and Strauss 1967, Becker 1998) aimed at revealing insights about boundary crossing activities, the sample consisted of thirty-three scientists who were members of the most diverse groups—those with representation from multiple academic departments. The sample was further refined through a bibliometric method developed to distinguish the most active boundary crossers.

Bibliometric studies are concerned with identifying patterns in the literature using indicators such as authors, textual content, citations (Borgman 1990) and acknowledgements (Cronin 1995). The literature is a primary vehicle for scientific knowledge accumulation. It is a kind of "form," a representation of something that can be collected and combined with "no limit to the cascade of re writing and re-representation. It is these new unexpected connections that explain why forms matter so much" (Latour 1987, 243). The force of texts lies in their connection to other texts. The literature provides a means for following researchers as they build links with other research work. Cited references document the information researchers receive, translate, and convert as they create a new scientific formulation for dissemination to others

(Callon, Law, and Rip 1986). As Latour points out, translation is different from diffusion. The concept of diffusion recognizes the movement of "facts" but not what happens to them as they are passed from person to person. Translation is best thought of as how scientists convince others that they have solved one of their research problems—the actions they take to turn their claims into facts.

Within the realm of bibliometrics, citation analysis is the technique that has been most commonly applied to examine the interaction between intellectual domains or to measure interdisciplinarity.[7] Scholars use bibliographic citations to connect themselves to the literature and subject areas that are important to them. Citation analysis assumes that citations have important symbolic value and that making reference to a text is a significant social act (Small 1978). Authors are mindful of who and what they cite, selecting references that come together to form an original composite that stands in relation to what others have done before. Citation studies contribute to our understanding of the "socioecology of scholarship" by exposing the combinations of resources used by scholars and the associations developed between subject areas as scholars synthesize new information in the context of their informed state (Sandstrom 1994).

While citation analysis succeeds at showing links and relations between literatures, it is not sufficient for understanding all the dynamics involved in knowledge creation. As Callon, Law, and Rip (1986) contend, it is not "an end in itself, but rather a method, albeit a vital one, for the pursuit of the qualitative by other means" (117). The technique is most useful for classifying and presenting statistical data to show "where to look" to gain a better understanding of knowledge development (Turner 1994). As such, in this study citations from the scientific literature were analyzed as the first step in identifying the real-world boundary crossing practices that cannot be discerned in the literature. Adopting Turner's (1994) scientometric approach, a technique was designed to build a foundation for the later qualitative stage aimed at revealing the activities and meaning behind the citation results. The citation analysis concentrated on a five-year period, 1989-1993, corresponding with the Center's first five years of operation. The method operationalized the notion of *scatter* to produce descriptions of the intellectual import and export of the scientists. Import was conceived of as the scatter of subject areas brought to bear on a scientist's work. Export was defined as the scatter of the scientists' work through dissemination channels. The results consisted of maps and profiles of the researchers' boundary crossing patterns. Details on the citation analysis methods and sample maps and profiles are provided in the appendix.

The import measure for each scientist was based on the references listed in their research publications.[8] Copies of the scientists' published papers were used to create textual data sets consisting of title words from each reference cited within the papers. The data sets were analyzed using Leximappe co-word analysis to create visual representations of the association between keywords. Co-word maps are thought to produce a truer reflection of the way science is practiced (Callon et al. 1983), relative to co-citation methods.[9] They illustrate the "trans-disciplinary" evolution of scientific production by showing the degree to which a problem area is internally structured by embedded concepts in relation to the development of new connections to non entrenched concepts (Callon, Law, and Rip 1986). Courtail and Law's (1989) co-word study of two research fields is a case in point. It showed that a more established research field (artificial intelligence) had an identifiable intellectual core, while a newer, less developed area (acidification research) had no core, just groups of themes that shared neither direction nor emphasis.

The structure of the co-word maps produced in this study showed the internal core and the emerging scatter of an individual scientist's research domain. It was then possible to identify the researchers with the weak links and emerging connections thought to facilitate the flow of information and the passage of ideas across disciplinary boundaries (Granovetter 1973, Cronin 1982). Scientists with loosely structured maps that displayed citation trails into multiple subject themes were considered active information importers. In contrast, the tightly structured maps with few emergent linkages were taken to signify low import, as shown in the sample co-word maps in the appendix.

In the next stage of the citation analysis, an export profile of each scientist's work was developed by assessing the scatter of citations *to* each researcher's publications. The profiles represented the potential breadth of audience for a given researcher's work. For each scientist, a citation table was produced that listed frequency counts for the journals that had cited their publications. The journals were broken down into the following categories:

- Multidisciplinary
- Disciplinary
- Subdisciplinary
- Interdisciplinary
- Problem-centered

The categories were based on established subject descriptions provided by the journals and informed by Klein's (1990, 1996) assessments of existing knowledge domains. The ratios of titles to items and subjects to titles were calculated, and the final evaluation of export, or dissemination scatter, was based on the *range of subjects* reflected by the journals citing an author, not the number of citations. Scientists cited in a wider range of subject areas were considered to be part of a cross-disciplinary discourse. In particular, cases showing more than one of the following characteristics were considered high information exporters:

- cited in science and social science citation databases
- cited in multidisciplinary journals with a large general audience
- concentration of citations in problem-centered or interdisciplinary journals.

In contrast, those cited primarily in general disciplinary and interdisciplinary journals were considered low exporters. Examples of variant profiles are provided in the appendix.

Relating ideas and evidence to a larger community is best achieved by using data with depth (Harper 1992) from informants who fit the conceptual requirements of the study (Glaser and Strauss 1967, Ellis 1993). To this end, the final sample was confined to the most active boundary crossers: twenty-two researchers who displayed high degrees of import or export. They either had co-word maps showing a loosely linked research terrain, as evidence of import, or they had diverse dissemination profiles, as evidence of export. Seven of the scientists measured high on both indicators.

The import and export measures were critical for identifying active boundary crossing scientists at the Center, but discovering how the scientists do this type of work required additional data collection and analysis. In the second, qualitative phase of the study, interview data was collected from the selected researchers. I was able to talk in person with all of the scientists except one, who submitted responses to the interview questions via e-mail. To gain a more complete understanding of the work environment at the Center, three additional participants were added to the pool: the Organizer and the Humanist who were introduced earlier in this chapter and a staff member from the computer systems office. The staff respondent provided knowledge of the computing and information technologies used by the scientists, and a complementary point of view on the interactions between groups and individuals at the Center.

The import and export profiles played a further role in this phase of the study as a tool for establishing rapport with the researchers. Respondents tend to be more forthcoming when they know that "you know a good deal about what is going on" (Becker 1970). To demonstrate to each scientist that I had an awareness of their work, I began the interview sessions by explaining the citation analysis techniques I had used and reviewing their co-word maps and export profiles with them. For the formal interview segment that followed, I worked from an interview guide consisting of about a dozen basic questions that prompted the scientists to tell their own stories about how they do interdisciplinary research. As Seidman (1991) explains, interviewing helps us to understand the experience of others and "the meaning they make of that experience" (3). The interviews were loosely structured to cover all the established interview questions while allowing researchers to talk freely about topics that were important to them as interdisciplinarians.

Analysis of the transcribed interviews was broadly based on Strauss and Corbin's (1990) model of grounded theory development, in which iterative cycles of data gathering and analysis were carried out to develop a thematic analytical framework (Glaser and Strauss 1967). All of the interviews were completely transcribed and all potentially relevant text was marked for analysis. The first several rounds of coding were dedicated to labeling and referencing the many things, activities, ideas and other elements represented within the text that might help explain the practices and conditions of the interdisciplinary research process. The terms used to describe the text were continually reviewed, categorized, and cross-categorized to uncover prominent themes and patterns, and the most substantial clusters of concepts eventually were used as titles of chapters and subsections in this book. Through this cross-case analysis all the data was brought to bear on the stated research questions. Single cases were also analyzed to explore how an individual's career path and memberships in various intellectual communities influence interdisciplinary research.

The accounts of interdisciplinary research that I provide throughout the book are largely based on the textual interview data. I have concentrated on relating research practices as they were portrayed by the researchers and on relating how interdisciplinary research is accomplished in the structures and cultures within and surrounding the Center. The respondents' words are used extensively to communicate their perspectives, perceptions, and experiences.[10]

THE SCIENTISTS AND THEIR WORK MODES

It is impossible to characterize the scientists as a uniform group, and it is also not possible to characterize any given scientist in any one way. As research faculty they achieved a Ph.D. in a program or discipline, hold a position in an academic department, and work in a field or problem area. To represent them by just one of these distinctions would be a misleading simplification of who they are and what they do. Less than half of the scientists have faculty appointments in the same academic discipline as their Ph.D., many are members of more than one academic unit, and almost all have strong affinities with teaching and research programs outside of their home departments. I have found that the researchers' own descriptions of their field of study provide the most useful information about where they sit in the intellectual landscape of academe. Figure 1 identifies each researcher with a title loosely based on their self-defined research area. The titles are meant to be specific enough to differentiate between cases, yet sufficiently general to maintain the anonymity of the individuals. For the same reason, gender representations have been altered for some cases while keeping similar proportions of men and women,[11] and names of colleagues given during the interviews have also been changed.

Microbiologist	Device Physicist
Structural Biologist	Condensed Matter Physicist
Bioenergetics Specialist	Perception Specialist
Protein Specialist	Human Factors Psychologist
Biophysicist	Movement Specialist
Biological Systems Specialist	Psycholinguist
Animal Learning Specialist	Language Modeler
Neuroscientist	Molecular Network Specialist
Computational Neuroscientist	Complex Systems Chemist
Neurophysiologist	Vision Specialist
Computational Designer	AI Specialist

Figure 1. Individual scientists by assigned title

Disciplinary categories have been considered dubious depictions of individual researchers and their intellectual communities for some time. As the anthropologist Clifford Geertz (1983) contends, even general groupings, like Natural Science, Biological Science, Social Science, and the Humanities, work to obscure what is "really happening out there where men and women are thinking about things and writing down what they think." The lines that separate and connect scholars run "at some highly eccentric angles" (6-7). The range of interactions between biological, physical, behavioral, and computational scientists observed in this study suggests that these lines may only converge within the conceptual space created by research problems. Who really talks to whom, through the scholarly and professional literature and through other formal and informal means, may have no other common factor. As one of the physicists explained, "the world doesn't know about physics, chemistry, and biology; the world's problems developed independently of them, so to solve them you really have to try to go at it from all angles." There is no one prescribed way to do this. The scientists at the Center have different ways of going at it from all angles, different modes for crossing boundaries to solve problems.

The scientists see themselves as working *on* problem*s*, not *in* disciplines or departments, yet they are not particularly comfortable calling what they do interdisciplinary or even multidisciplinary. This problem orientation is not uncommon in industrial or medical research settings, but, as demonstrated by the need for specialized interdisciplinary institutes like the Center, it is not typical in the academic world. Nonetheless, problem-centered research at universities has been advocated for decades (Long 1971), and research teams working between the poles of pure theory and informed action have long seen themselves as "problem-focused" (Klein 1990). My application of the term "problem-centered" incorporates the various types of boundary crossing that occur in scientific research, including movement between domains and between theory, experimentation, and application.

Researchers at the Center understand that scientists and engineers need a different approach than in the past if they wish to succeed in cutting edge, hybrid research areas. As one scientist explained, it is no longer enough to know how to "find the engineering solution for the piston." The problem must be understood at all levels of its life cycle, from the "design elements of the engine to business and marketing." His concerns echo Thomas Kuhn (1962), who asserted that members of research teams must grasp the overall problem and that multidisciplinary teamwork requires "much more broadly educated researchers" (Harwood 1993, 235). But the scientists at the Center have not been part of a formal program of broad scientific training. As one biophysicist pointed out, little has been done on a conscious level to educate scientists to be interdisciplinary. Researchers at the Center have crafted their own interdisciplinary modes of work through shifts in research focus, educational pursuits, and career paths, and occasionally, by working with interdisciplinary mentors. Three primary modes of boundary crossing research were identified among the scientists studied: collaborator, team leader, and generalist.[12]

My method and purpose for developing the research modes parallels John Law's (1994) conception of "modes of ordering" in his study of laboratory managers. He uncovers four patterns in his data—enterprise, administration, vision, and vocation—and demonstrates how the categories are recursive and interactive. Like Law's modes of management, the boundary crossing modes I have developed are "ideal types," not "pools of total order," or the only possible types. Instead, they should be thought of as dominant orientations or preferred ways of doing research. In practice all the researchers are multi modal to some degree, a recognized feature of interdisciplinary inquiry (Klein 1990). As Law explains, such categories provide "tools for sensemaking," not a rigid framework.

Over time, the scientists have adopted and applied methods and ideas from multiple scientific communities. Many have taken on a wider perspective in their work, increasing the scope of the questions they ask and expanding their information domains accordingly. While a given researcher will generally favor a particular mode, they also position themselves strategically depending on the problems they are addressing, at times changing or working across modes. Nevertheless, the modes are grounded in the full set of data and are useful constructs for understanding how boundary crossing research work is accomplished.

RESEARCH MODE	**Collaborator**	**Team Leader**	**Generalist**
APPROACH	cooperative	managerial	individualistic
INFORMATION PRACTICES	finding	gathering	probing
KNOWLEDGE STRATEGIES	consulting	recruiting	learning
SCOPE	depth	breadth	breadth
OUTCOME	productive	productive	integrative

Figure 2. Interdisciplinary research mode characteristics

The five categories of characteristics that distinguish the three modes are outlined in Figure 2: approach, information practices, knowledge strategies, scope, and outcome. *Approach* is the most basic feature of the modes, in essence a descriptive term closely aligned with the mode name. It is the general way research work is administered or carried out within a mode. The scientists tend to be cooperative, managerial, or individualistic in their general approach to solving research problems. The remaining categories of characteristics will be addressed in detail in the chapters to come. To help assimilate the characteristics into the conceptual modes, I have highlighted exemplar cases for each of the three modes. These three scientists, introduced below, will direct the narrative or tell much of the story, so to speak. Together, their perceptions and experiences encompass the spectrum of practices and issues reported by the entire group researchers. They are well established scientists who have strong research records and years of experience developing their areas of expertise and interdisciplinary practices. Examples from the remaining cases are woven in throughout the text to support and supplement the themes brought to the forefront by these three archetypal figures.

Collaborators

Collaborators prefer to address research problems by working with colleagues from other domains. Tom Tate, a Complex Systems Chemist, represents the collaborator mode. He learned the value of cooperative science early in his academic career.

> When I was in both college and graduate school I thought the purpose of life was so you could become the universal knower of everything, and certainly to know more than the next guy over, so that you could accumulate resources and do whatever you wanted to do because you knew more than the other guy. And that's a losing game. So then the next step is, well, what do I know that I can actually contribute to some cause? I then join up with somebody else who knows things I don't know and doesn't know the things I do know, so together we can do something that no other guys do. That's pretty much the way I function.

Tom believes that a participatory approach is the best way to be a "useful citizen" and to make real progress. Collaboration is a widely practiced interdisciplinary technique, and the way it is carried out, the purpose it serves, and the level of exchange between collaborators takes many different forms. In Chapter 4 we will examine the various ways Tom collaborates, which may be consultative or integrative in nature depending on the problem at hand and the collaborators involved.

In addition to his faculty positions in chemistry, Tom has worked in the private sector and for national agencies. He has done extensive work on instrumentation for chemical analysis, but his connection with the Center is through his more current interest in complex systems.

> I went to a presentation on nonlinear systems where all of a sudden I said: that looks really neat; I've got to learn more about this. Right about then it became clear that I couldn't just do one narrow thing and accomplish anything. I had to pick up stuff from other people, and they weren't about to waste their time teaching me about what they knew unless I had something to teach them.

Tom's new projects have required him to branch out into biology, biochemistry, physics, and cybernetics. Because of this recent transition, Tom's case provides valuable contrasting accounts of disciplinary and cross-disciplinary information gathering and cooperative work scenarios.

Team Leaders

The team leader designation describes the scientists who talk about themselves primarily as managers and who oversee groups of researchers working on multiple projects. Sam Smith, a Biological Systems Specialist, represents the team leader research mode. Sam has studied and worked at prestigious universities and previously held a position at another interdisciplinary institute. He was attracted to his current situation by the opportunities afforded by the Center, where he now manages a large laboratory operation made up of mostly graduate students and postdoctoral fellows. As a team leader, Sam plays the role of visionary and guide.

> When there is a team of several people contributing they very quickly lose sight of what was the goal and that they really have to do everything to optimize achieving the goal—not making one solution really good and another so poor, and because of some poor subsolution everything will crumble. Like in a rocket, everything must work equally well or the rocket might explode.

Sam implements diversity, and in some respects, acts as an enforcer of interdisciplinary exchange. He has an acute awareness of the confines of scientific disciplines, and works to merge "different cultures" and "different attitudes." For example, he places great value on the presence of a computer scientist in his lab that has extensive experience in the private sector.

> This fellow has been working with the rigor and discipline and careful documentation of everything you have in industry. The graduate students from physics, who all think they are little Einsteins in the lab, first they see the other guy as really odd. But they are all nice, intelligent people; they appreciate each other's abilities, and they agree together on a goal.

The computer scientist enriches the group with his knowledge and skills, but also by creating a situation where the "little Einsteins" must interact with a little Babbage. They learn to work together and, as Sam points out, not to "overvalue their own disciplinary background."

Generalists

Interdisciplinary science is often equated with collaboration, and indeed cooperative and team research abounds at the Center. But, since existing groups of researchers were targeted in the initial recruitment of scientists, the level of interdisciplinary work generally performed by individuals is likely to be underrepresented at the Center. Independent boundary crossers practice in a Generalist mode. They often work alone and tend to build a strong personal base of knowledge to address broad research problems. Neil Nash, a Bioenergetics Specialist, represents the generalists.

> I didn't actually decide what I wanted to specialize in until the last minute, and so I had to have fairly broad interests as a biochemist. . . . In a way, I think my breadth might have actually been a reaction against a perception. Even as a student I was aware that scientists were narrow.

Neil is an accomplished biochemist who has been working at the University for almost twenty years. He has held demanding administrative positions and been recognized with awards and fellowships throughout his career. He rarely collaborates with others in his research, and this perhaps explains why the Center has not been particularly important in nurturing his work. Instead, his research approach has become more integrative as his field has grown more multidisciplinary on an international scale.

While the researchers working in the other two modes were based in a variety of academic disciplines, the generalists clustered in biophysics and psychology, fields where disciplinary boundaries have already been dismantled to some extent. A number of the generalists became more interested in theory and modeling after years of experimental work. As one generalist explained, "I came to the point where I was data rich and concept poor." Neil avoided this difficulty by learning how researchers in other areas address problems and work with similar data.

> Once you've got the data, understanding data can be done at many different levels. We may have a very biological or biochemical view of how something works. For example, we may be interested in how bits of the protein wiggle about, and how the environment changes the shape of the protein, and how that shape change may affect some ultimate function. But the person coming from the chemical physics background may be interested, and probably is interested, in how those changes in shape and environment affect the electronic spectrum, and in effect, the quantum physics of the process. So both sides actually gain questions as well as answers and insights from all of the different perspectives.

With their less specialized orientation, Neil and the generalists emphasized different dimensions of the research process than the collaborators and team leaders. Neil reflected at length on the role of computation in broadly based theoretical work and was especially conversant on the communication barriers between scientific groups.

> This is Deep Space Nine[13] you know, and you have to really be able to communicate to both ends of the spectrum. Not the furthest ends, but

> you have to be able to present to them a fairly broad level . . . from the small molecule right up to the integrated biological process.

SENSE OF PLACE

As we consider the modes that researchers use to solve problems, we must also take into account the context of research activities. The structures in which researchers conduct their work and the conditions in those places are part of the accumulation process. As Fisher (1990) notes, although individuals carry out boundary work, it simultaneously involves institutions and the social structures that envelop people and their work. Research organizations function as centers of translation. They organize meetings, symposia, and study sessions; they orchestrate accumulation through the "circulation of inscriptions, the movement of people, materials, and money" (Callon, Law, and Rip 1986, 27). Members of the Center work in multiple structures and contexts, which creates complications but also possibilities for successful boundary crossing. As they move between places and change roles, the scope of their contacts and experiences expands, creating more options and opportunities for exchanging information. Interdisciplinary researchers configure their memberships and places of practice for the purpose of boundary crossing accumulation. They belong to multiple groups; some are stable and institutionalized in university departments and professional societies, and others are emergent informal networks of cross-disciplinary colleagues.

Some of the scientists consider the Center their primary place of research; others reside there on a part-time basis. Of the three modal representatives introduced above, Sam is the only one whose research activities are seated solely at the Center. Tom and Neil have continued to work primarily out of their departmental labs. "Place" is a complicated issue for many of the researchers, both physically and figuratively. They work in more than one place, and for some no place feels like home. They are situated in a number of different scientific communities, but may not be a recognized member in all of them. One very active boundary crosser laughed about his many attachments. Locally, he has a place in the physiology department, the biophysics program, the bioengineering program, the neurosciences program, a computing research unit, and the Center. And this list does not include his national and international group memberships. Some researchers use their Center or departmental offices and labs as space for their factions to intersect. Others linger on the fringe of departmental and disciplinary territories, struggling to find or make a place for their work.

Emphasizing the social and political factors involved in the research process, Lenoir (1993) makes an important distinction between the research programs undertaken by scientists and the organizations that support them. Scientists develop programs of research to help define their scientific field and to extend and legitimate the products of their work. On the other hand, the organizations that house and manage these programs are a base from which disciplines are built and strengthened, enabling the transmission of techniques, tools, and information within an institutional niche. The differentiation between the research program and the research organization is important for understanding scientific work at the Center, where each scientist is associated with at least two—often conflicting—niche-building institutions. Each researcher is a faculty member in an academic department that works to strengthen its presence in the scheme of established disciplines. They are also members of the Center, a unique administrative unit constructed to advance scientific research fronts and build hybrid disciplines. Their work environment, made up of both discipline- and hybrid-building agencies, necessarily compounds the socio-political tensions surrounding research.

In addition to the Center and the academic departments, there are other cross-disciplinary frameworks within the university that impose further impacts. They stem from the less formal teaching and research affiliations that transpire on campus. The "biophysics faculty" is one such group. It has a loosely organized membership that includes individuals from biology, chemistry, biochemistry, physics, agriculture, and the medical school. A number of the biophysics faculty have appointments at the Center, but continue to work exclusively out of their departmental offices and labs. As the Center's initial selection process capitalized on and centralized cross-disciplinary groups that already existed on campus, it also aimed to unite the existing groups of biophysicists and neuroscientists. The founders wanted research groups to undertake problem solving at a higher gauge.

> We are trying to boil down these groups together in the same building, which is sort of a collection of interdisciplinary centers. The hope was that there would be some longer range connections made. (The Organizer)

Once they became members of the Center, the researchers were expected to move beyond the projects formulated by their self-organized groups. In this sense, as one scientist suggested, it was to be a "hyper-interdisciplinary" research institute. "I am already interdisciplinary and now I am being asked to be even more interdisciplinary by mingling with folks from engineering and all sorts." As a whole, the Center might be thought of as one big laboratory equipped for large-scale participatory science. Cross-group interaction is a fundamental goal, and, as the Organizer explained, it is "one big experiment." Functioning as a lab of labs, or a meta-laboratory, the cogged units should, over time, have an impact on the entire organization.

After more than five years in operation, some of the Center's research groups were moving in concert, while others had not meshed and remained isolated. This could be due to the newness of the organization. Perhaps, as the Humanist suggested, some are still "warming up" for the performance, or certain laboratories may be conducting their own separate orchestra. For example, Sam's lab is addressing a broad range of problems internally, but there was little evidence of external involvement with other parts of the Center. Nonetheless, the group is anything but isolated. Through his students, Sam is connected to numerous campus units outside the Center, and clearly has the attention of many other scientists at the university and across the world. In fact, Sam's student workforce is a prime example of the successful scientific apprentice system admired by the Humanist.

The groups and individuals affiliated with the Center work in different ways to accumulate the knowledge needed to solve their research problems. With all the plans for the space and the people realized, the Center acts as a defined place for the yet undefined practice of interdisciplinary science.

---

[1] For the most comprehensive body of work see Klein's (1990, 1996) two major integrative studies. Her extensive bibliographies in themselves stand as the consummate record of research and scholarship on interdisciplinarity.

[2] Quoted from an informational brochure produced by the Center's external affairs department.

[3] Industry has had a longstanding concern with the design and construction of productive research environments. See for example Carlson's (1991) study of Elihu Thomson at General Electric between 1870-1900.

[4] Fisher develops his concept of integration from Bernstein's (1977, 1982) work on integrated educational codes.

[5] For example, in a review of "yet another" festschrift honoring the life and work of Dobzhansky, Jerry A. Coyne (1995) comments on how "ancestor awareness" has become a form of "ancestor worship" in the field of evolutionary biology.

[6] See Clarke and Fujimura (1992) for a comprehensive review of influential works on the situated technical culture and practice of science, and Becker (1982) for a similar approach to the "sociology of occupations applied to artistic work."

[7] See, for example, Chubin, Porter, and Rossini 1986, McCain 1986, Choi 1988, and Hurd 1992.

[8] The import measure was based on a sample of 50% of each researcher's publications identified from their recent vitae and by searching the Institute for Scientific Information's SciSearch and Social SciSearch databases to find additional recent items. Works that had multiple authors were included since collaboration is a common feature of interdisciplinary research. Technical reports issued by the Center were also included in the analysis process.

[9] See, in particular, Small 1977 and Small and Greenlee 1980.

[10] English was a second language for a number of the researchers interviewed. On occasion I have slightly altered their sentences to improve readability, but for the most part I have reproduced quotations as they appeared in the interview transcripts.

[11] The gender ratio for the sample differs slightly from the .11 ratio for the entire Center membership.

[12] In a study of biochemists, entomologists, and statisticians, Judith Palmer (1991) identified five information styles, three of which roughly correspond with these modes. The characteristics she attributes to "lone-wide rangers" are generally in line with the information practices associated with the generalists. Both the team leader and collaborator modes seem to be congruent with her "confident collector" category.

[13] This reference is to *Star Trek: Deep Space Nine*, a science fiction television series about a remote station at the edge of a new frontier in outer space. "Travelers of all kinds are drawn here, and with hostile alien empires bordering every side, Deep Space Nine becomes the most strategic point in the galaxy." (Description from a Paramount Television Current Productions web page, July 2001.)

Chapter 2.

INFORMATION ACCUMULATION

Information work is a means to an end. Researchers seek, gather, mobilize, and synthesize information to work on research problems and produce new knowledge. In the field of information science, we inform the design of information systems that support research work by studying the products and practices of researchers. We learn how to improve the organization of information and the functionality of information technologies by observing how information is gathered and used. At the same time, the patterns of information import and export observed reveal how knowledge moves and coalesces among scholars and their intellectual communities.

For the purposes of the first phase of the study it was necessary to delimit the definition of import and export as reflected in the researchers' citation practices. The more complete view how information is imported and exported across boundaries—the many ways it is accomplished, reasons why it is done, and the merits of various strategies—emerged from the interviews conducted in the second phase of data collection. Information work is complex and demanding for researchers who work across multiple domains, and for many at the Center it is less routine than it used to be. As Tom Tate notes, he can no longer go to the library to see everything he needs for the week. In this chapter, I describe the import and export practices used by the researchers, highlighting the techniques that were emphasized by the scientists during the interviews and those that appear to have been most successful. As a group the researchers consistently talked more about the import of information than export. They were very conversant about how they get what they need to solve the problems they are working on, but few talked at any length about their dissemination practices. In a later chapter I will come back to this imbalance between import and export in the research process to consider its significance for scholarly communication and the role of information systems and services more generally.

To describe the information work of researchers, I have made a distinction between processes and practices. Concepts that describe ongoing import and export processes are listed in the top portion of Figure 3. A repertoire of practices that are used in combination to carry out these processes is provided in the lower segment of the figure. The researchers specifically discussed all the practices listed, and the processes were widely demonstrated across the group but not necessarily distinguished in the scientists' self reports of their information work. It is important to keep in mind that import and export are interrelated within the accumulation process, although the researchers do not tend to talk about their information activities in this way. Import and export are necessarily related in that one person's import is another person's export, and export activities frequently open up channels for future import of information. Moreover, not all tasks belong exclusively in one category or the other. Researchers who collaborate, trade, and correspond are participating in an *exchange* of information, therefore exchange-based activities are listed in both the import and export columns in Figure 3.

Processes	
Import	**Export**
seeking	targeting
gathering	generalizing
probing	fitting
learning	
Practices	
Import	**Export**
attend	analogize
bridge	bridge
browse	collaborate
chase	correspond
collaborate	customize
consult	discuss
correspond	present
discuss	publish
network	reference
read	rework
retool	rewrite
referee	synthesize
skim	teach
teach	trade
trade	

Figure 3. Import and export processes and practices

THE MEANS OF IMPORT

The practices scientists use to import information are subsumed in varying combinations within the processes of seeking, gathering, probing, and learning. Information *seeking* is a hunting effort directed at finding a certain type of content or specific items. For instance, when Tom Tate is looking for information related to engineering, he goes directly to a colleague. "I just use him. I don't have to do anything else." Information *gathering* is less directed and more loosely organized around a problem area. Researchers gather when they perform a literature search on a topic, and it is the primary means for keeping current. Information seeking and gathering activities have long been associated with academic research, and they were part of every scientist's information routines to some extent. Team leaders emphasized organized techniques for group information gathering and exchange, while collaborators stressed the need to find specific information for their cooperative projects. In interdisciplinary work these tasks are complicated by the need to venture into multiple and often unfamiliar territories. Keeping up with and using information across fields requires maintaining awareness of the many different facets of a research problem, the new work in relevant subjects, as well as potential emerging areas of importance. Interdisciplinary gathering is problem-centered, drawing on diffuse areas of study that may coalesce over time.

New applicable work is often identified through information *probing* activities. Probing is similar to the practice of surveying, as described by Ellis (1993); it is a way of becoming familiar with the literature or body of knowledge on a topic. Probing differs from surveying in that it is more investigative in nature. Researchers probe broadly into peripheral areas to increase their breadth of perspective and to generate new ideas. Probing activities may be either sweeping or directed. An optics specialist might skim through a wide range of journals, general science magazines, and popular scientific texts hoping to latch onto insights in subjects ranging from psychophysics to graphics. Researchers also probe deeply to gain an understanding of how related fields can be applied to their work. For instance, one psychologist attended intensive workshops outside his core discipline for this purpose.

Information seeking and gathering tend to be continual accumulation processes. All the researchers were concerned about the problems they experienced seeking and gathering, and not surprisingly they had particular difficulty locating material outside their core specialization. In contrast, probing was not discussed as a problem but as an important information strategy. Information probing can lead to the broadening of a research area, and the infusion of new information or knowledge can lead to a shift in a researcher's problem focus. This type of change is analogous to how a query evolves through the information search process, as described by Bates (1989), where each new piece of information encountered can change ideas, directions, or the essence of the original search. In contrast, the changes in direction and ideas that result from probing are largely intentional. And, as outside information is assimilated, the realm of relevant subjects to search and keep current in is altered and, in most cases, expanded. Furthermore, with each new domain there are new terms and concepts, background knowledge, and analytical approaches to learn. Overall, the three processes of seeking, gathering, and probing are dynamic and interconnected.

Interdisciplinary researchers have different information needs and strategies than disciplinary researchers, but their general sources of information are much the same. The role of formal and informal communication has been documented in various scientific contexts. Garvey and Griffith (1964) found that literature and conversations with colleagues are emphasized in different stages of the research process. In a comparative analysis of scientific information use studies, Skelton (1973) found citations, abstracting and indexing services, personal recommendations, and chance to be the most important methods for retrieving information. Likewise, the Center researchers use both formal and informal information channels for importing information. They face challenges accessing and using these sources and must make decisions and set priorities that will help them overcome the barriers of working in new or unfamiliar knowledge domains. They need to make sure that their time and effort are spent targeting the relevant publications and making the right contacts in outside fields. None of the researchers was at ease with the boundary-crossing import process; it was a practical and intellectual challenge for all. Experienced scientists feel like novices as they look for information in unknown surroundings and attempt to become oriented to new research contexts. As the researchers talked about how and why they import information, they concentrated on personal networks, conferences, and the published literature. They also emphasized materials and activities that helped them learn something new and the role of intermediaries as a pivotal component of the accumulation process.

Personal Networks

> Normally, maybe 85% of what's going on I just know by keeping in contact with people and by going to our own conferences. (The Device Physicist)[1]

For many researchers at the Center, personal networks are the most important vehicles for information exchange. Networks of colleagues and students are rich sources of information. Because they are efficient and yield quality results, they are relied upon for keeping current and for consultation. This is consistent with early studies of informal scientific communication that established the importance of conversations and correspondence for the exchange of news and for getting feedback on preliminary work (Garvey and Griffith 1968, Griffith and Miller 1970). In interdisciplinary research, feedback from knowledgeable sources is crucial because of the uncertainty involved when venturing into unfamiliar domains.[2] Furthermore, personal contacts are one of the few defenses against missed literature, a common and potentially serious import problem. Conversing with people in allied fields makes researchers aware of their knowledge gaps.

> I'm perpetually finding that no matter how much time I spend looking for things somebody will walk into this office, from name your favorite subfield, and say, oh have you looked at—and they will name an article that is smack in the center of something that I care about, and I've missed it. (Tom Tate)

Establishing contacts in other fields promotes understanding and integration across fields. Researchers consult with contacts from different backgrounds to explore the various ways a problem can be approached, to grasp the long-term hopes for a solution, and to learn how their research relates to other work on the topic. The exchanges that take place in these multidisciplinary networks are small steps toward scientific convergence.[3] The Vision Specialist used the metaphor of huge interactive databases to illustrate how people from different disciplines home in on relevant information about a problem.

> Someone will say, oh this guy did something, so and so, you should look at this paper. This will happen after they have summarized the significance of that, which they did not know until I told them what I was looking for. So it's an interactive search for the right thing. They have their own huge database, and if we talk, then it's converging on the right references or people. Or even if they refer you to somebody, they would know in a sentence what is there; they would already say why before they point me to the reference. They would say what that is, and so if that's relevant then I would go. And of course it is likely to be relevant, since they already know what will be relevant.

The interactive process narrows the discussion to a specific concern and centers it within the perspective that is needed. No other type of information interaction is likely to be this efficient and profitable. The members of a personal network function as filters and interpreters.[4] Networks are ideal information filters because they are personally customized to an individual's needs and activities. Interpersonal connections are established through shared interests and tend to be made with trusted colleagues who have the authority to help evaluate information.

> I'm not going to take a lot of time out to say; well I'm going to struggle with this to see whether there is anything in it. I've got to wait for someone to say, well you know, this new formula really looks promising. If someone can tell me whose judgment I value, then I'll say, okay, fill me in a little bit. Then I'll talk to them and then I'll find out, and then maybe I'll go off and start doing some work on it myself. (Neil Nash)

Researchers who do a lot of information probing are frequently faced with the task of sifting and evaluating all the ideas and "pet theories" that they encounter. Personal contacts from outside fields are called on to help determine the viability of newly discovered material from outside domains. Even researchers like Neil, who prefer to read about emerging areas of interest, almost always follow up by discussing particulars with colleagues. For example, he established relationships with scientists who could help him cope with the physics he needed for his research.

> I'd find the people who are from a more theoretically sound background, physical background, and then I would ask them, how are you handling this? How do you interpret this? What's the pitch of the mechanism you are seeing? They are the people who are most likely to be able to interpret it for you.

Talking with someone in the other field helps him assimilate the information gained through the literature. Together they turn that information into usable knowledge.

The value of e-mail for managing the exchange of information within personal networks cannot be overstated. All but one of the researchers talked about e-mail as if it were an indispensable part of the research process. It is used "perpetually" as the primary means for keeping in contact with colleagues. It has made a significant difference for two activities in particular: planning and collaborating. It is how the scientists "get organized, arrange to do this and that." For example, the Animal Learning Specialist found e-mail to be the ideal tool for managing his editing projects. It allowed him to keep in touch with many authors simultaneously and compile and edit the texts at a pace that suited his schedule.[5] Other important features of electronic communication, addressed later in the text, relate to its expediency for cooperative work.

Conferences

> If it weren't for conferences I really would be lost. (The Photosynthesis Specialist)

> Everybody who matters is there and for a week you get saturated in this stuff. For what I am doing, I have to be at that conference. (Tom Tate)

Some conferences are essential and some are marginal. The researchers had limited interest in large, general meetings, and they talked more about conferences in terms of import than export of information, even though their vitae showed that they are frequent presenters at such meetings.[6] When the Device Physicist, quoted at the beginning of the last section, indicated the importance of "our own conferences," he was talking about a fairly small meeting of a national group of scientists in computational electronics. The researchers rarely mentioned large discipline oriented conferences. The most valued meetings are those that congregate at the problem level, where researchers feel part of a "closely knit group" that shares specific research interests. [7] Small, specialized meetings were considered by many to be as critical as personal contacts for keeping up with information. In fact, it seemed that for some researchers these meetings are an extension or manifestation of their personal network. Like networks, the conferences satisfy a multitude of information needs. According to the Neurophysiologist, "you kill a lot of birds with one stone; you get the social interaction, you get the professional interaction, and you get the references [to the literature]." The meetings provide the efficiency, focus, and interpersonal aspects of personal networks held physically captive for days. In addition, this framework of intensified exchange is an ideal setting for establishing new connections with people who can improve or refresh your personal network.

Contacts made at specialized meetings may become research partners. Finding collaborators is a natural part of the act of assembling and talking about research. Neil suggests:

> The way it happens is by finding the people just in the normal sort of processes of social intercourse at meetings. You find the people who are talking in a way which you have some affinity for, the people who

> are making an effort. And then you talk to them and after a while you get to know them well enough that you can ask them stupid questions without feeling really idiotic. And really, there are some people who turn out to be just absolutely wonderful expositors of complex ideas—people who themselves have thought about, you know, why am I doing this. And you latch onto those people.

As a generalist, Neil seeks colleagues with the ability to communicate about research on a broad, reflective level. Those working in the collaborative mode also use conference gatherings to identify potential project partners, but they are more likely to seek out those with a specialized expertise or skill.

The more diverse the population at an event, the more cross-cultural issues come into play. Neil described one of his regular conferences by once again invoking *Star Trek Deep Space Nine*. "There are all sorts of different species around, some of whom can't talk to each other, no doubt about that." Neil has seen a tremendous influx of theoretical physics, computational studies, and both theoretical and experimental chemistry into biological protein research. It has become a "real melting pot." Overall, the communication difficulties encountered at conferences seem to be much less frustrating than those faced in the literature. As with other person-to-person information activities, the element of interchange is what makes it productive and satisfying.

The kind of information that can be picked up at a conference is quite different from what appears in the published literature. The Computational Neuroscientist referred to the conference experience as getting in touch with the "undercurrent." He found the information from discussions at conferences most valuable because it is "raw, not polished—because it is speculative," with no deep ideas attached to it yet. He particularly likes sitting around and talking about pieces of data that people don't quite know how to put together yet. This type of exchange offers a kind of intellectual stimulation that can't be found in the seamless research reports required for journal publication. Earlier studies suggest that this search for the undercurrent is also a goal of formal, disciplinary conference networking. The classic American Psychological Association studies (1963) on scientific information indicated that 80-90% of useful information at conferences was gained by attending formal presentations and events. A study of interdisciplinary behavioral scientists conducted around the same time period reported comparable results (Paisley and Parker 1967). Compton (1966) may have tapped the advantages of behind-the-scenes conference information when she found that attendees received a substantial amount of useful, but "unsought," information.

Earlier studies of scientific communication show that reliance on interpersonal networks is nothing new, and in many respects, interdisciplinary interpersonal communication and conference activities mirror what is happening in discipline-based invisible colleges (Crane 1972). In a synopsis of invisible college research, Cronin (1982) observed that the informal systems enable scientific communities to manage their affairs at a number of levels. They establish priorities for investigation, have a bonding effect on groups, support current awareness, allow researchers to screen information, and promote "boundary spanning," by transmitting ideas across disciplines (224). In the 1970 Johns Hopkins studies of science communication structures, Lin, Garvey, and Nelson found that 54% of conference attendees made contacts with others in order to exchange information, and 30% percent of those contacts were with new or unfamiliar researchers. Garvey and Griffith (1964) suggest that the greatest incentive for initiating new contacts may be to break into the invisible colleges that tend to be controlled by conference presenters. The researchers at the Center did talk about how hard it can be to gain entrance into established groups, but the problem seemed to be tied to a lack of a common knowledge base and vocabulary as much as the power group relations associated with invisible colleges. Price's (1961) conception of an invisible college—the "ingroup" in a field that controls personal prestige and the fate of new scientific ideas—does not readily fit what seems to be happening in hybrid fields of study in early phases of development.

Literature

> If you can't look at far more literature than anyone has time to look at you get into this tiny little corner where you keep reinforcing your preconceived notions. (Tom Tate)

Literature is a primary component of the accumulation cycle. As discussed in Chapter 1, it is a form of knowledge representation that can be collected and recombined, and there is no limit to the reworking of texts (Latour 1987; Callon, Law, and Rip 1986). Texts are standard resources for building knowledge, but finding and using texts in an interdisciplinary field is not a straightforward process. Tom's statement above brings together two of the greatest literature problems in interdisciplinary work. First of all, there is much more to read than anyone can possibly keep up with; the sheer magnitude of potentially relevant material seems insurmountable. Overload, as a consequence of interdisciplinary inquiry, is discussed in more detail in Chapter 4. Scatter is the other distinctive feature of the literature that can be brought to bear on interdisciplinary research problems. Literature dispersion has become a difficulty for Tom since he shifted from a specialization in chemistry to complex systems research.

> There was a time not that long ago when I could go to the physics library and walk from one end of the shelves to the other and inside of a half hour see everything I needed, and be pretty sure I hit everything that mattered, because I knew what journals it was going to be in.

In the past, searching electronic databases had been productive as well, because his interests could be covered in "only about ten keywords." There are so many sources and terms that relate to his current problem area that his old reading and searching routines are no longer adequate. Tom also commented that his increased subject scope makes it too expensive to have literature searching done by an information service provider. Moreover, the Internet has complicated the dispersion problem for Tom.

> It doesn't matter how marvelous the stuff is that is out there if you can't get at it, except if somebody says, by the way, I was talking to a guy when I was at a conference last week and he says that if you go on to this computer here you can find an address to go to that computer over there, which supposedly will tell you another place over there where you can get what you want. Now what kind of nonsense is that?

Tom compared the chemistry information available through the Internet to the state of chemistry literature before *Chemical Abstracts* began. His universe of literature had increased in two ways, almost simultaneously. The scope of subject areas related to complex systems is expansive, and, as a relatively new field for Tom, he had not yet developed a system of cues or filters to help him navigate in that domain. Added to that was the growth of sources appearing on the Internet, which at the time was the part of the information environment that seemed the most out of control to him.

In spite of his frustrations, Tom was one of a minority of researchers who emphasized the importance of electronic networks for functions other than e-mail. This situation has likely changed for many of the scientists, but at the time most did not generally include electronic formats in their conception of literature sources. The subject rarely came up as the researchers discussed their literature use, and when I asked specific questions about it, the responses were very limited. Many simply commented that they "should" or "wished" they could take advantage of available technologies for finding or exchanging information and documents.[8] Since that time, databases and full-text online journals have become more widely available through university library systems and more frequently used locally and nationally.

Despite the problems identifying and accessing dispersed literature, the researchers continued to read on a regular basis. The co-word maps developed in Phase 1 provided a picture of the content and breadth of the literature associated with each case, and such quantitative methods are valuable for identifying interdisciplinary relationships. Citations are a good indicator of contact with or awareness of literature. At the same time, they are artificial representations in that they treat each bibliographic item equally and don't provide insights into the actual act of reading, the intellectual application, or value of the material cited. The interview data filled in the gaps by adding details about ways of reading and the role of literature in the research process. In particular, they demonstrated that the act of reading is highly variable. There were clear differences in how and why the scientists read. For example, collaborators tended to be selective, highly focused readers.

> I don't go and read articles unless there is a specific reason. I don't have time to go and read other people's articles. (The Computational Designer)

> There's very little time for reading, so one has to rely on sure references before spending time and going through a paper carefully. (The Device Physicist)

The Molecular Network Specialist made similar comments, but, as a theoretician, he was speaking specifically about his interaction with experimental literature.

> I don't read [about experiments]. I would like to, but I don't. So in those terms I am not a good practicing scientist. . . . It is relatively inefficient for me to spend time doing that.

Like many of the other generalists, he preferred a different sort of reading that is broadly based and geared toward infusion of new and more generalizable knowledge.

Neil used to read regularly before he took on his current administrative duties:

> What I read in the literature, I mean the research literature—I think was generally much broader than most people in my area. And I think that really did help feed into—I mean, it gave me knowledge of just the way proteins in general function rather than keeping me very focused, and rather narrowly so, on what was being considered by the central part of my field. And I think that it did give me some ideas as to what might be happening that wouldn't have occurred to me otherwise.

Here reading is thought of as a strategy, a way of keeping an interdisciplinary edge. Broad reading is an important technique for sustaining a wide perspective, developing interests, and opening "broader vistas." Variations in reading patterns correspond with the boundary crossing modes and their approaches to knowledge development, a significant part of the interdisciplinary research process addressed in Chapter 3.

Researchers who read broadly, however, do not necessarily read carefully. Eighty-three percent of the respondents to the general survey of Center members at large tended to skim literature instead of read it. In the interviews, some recalled a time when they had been able to read entire articles and some journals cover to cover on a regular basis. Now, documents are usually skimmed, and whole bodies of literature are browsed. According to Chang and Rice (1993), browsing is searching that can be "both goal directed and nongoal directed and unplanned rather than aimless." These scientists, like most researchers, practice goal directed browsing to gather information, but they also probe new and peripheral areas, a practice that could appear nongoal directed or unplanned. For example, one researcher commented that he had found significant works by browsing contents pages at bookstores and at publisher's displays at conferences. Runs of journals and indexes and abstracts are also browsed. In both of these cases, there is a deliberate goal or objective to the browsing—a discovery or an intellectual leap of some kind. As precious as time is to these scientists, the potential for discovery is great enough that browsing is worthwhile. Some browse vast amounts of literature hoping to "trip over something by accident"—a reference, a mention of an idea, or a vein of thinking that might be important to their work. Tom's advice is to not "narrow things too much in the hopes that you are going to catch something. You've got to keep that peripheral vision up."

Broad conceptual and "summary books" that take a comprehensive view of science are important to some of the generalists for the insights they provide. General and comparative journals are favored for probing as well. Multidisciplinary titles like *Science* and *Nature* were cited as regular browsing and reading material. The Molecular Network Specialist read *Scientific American* religiously because it enabled him to "dip into things like software design and immunology," things he only had "a smattering of knowledge about." Then, once the vistas are opened, it is time to "put on your boots and slog through the literature." Once researchers move outside their core literature, it feels more like slogging than reading because the content and terminology are unfamiliar and cumbersome. A specialist in an area can easily skim titles and abstracts; a novice will need to spend more time and read deeper to determine what information is pertinent.

Some researchers browse electronically, but in general, the electronic databases were relied on more for finding information about something specific. Only a few researchers placed much importance on the online bibliographic databases available at the campus libraries and through the campus computer network. In the general survey, 43% of respondents never used the electronic abstracts or indexes available in the libraries, and 62% never used electronic versions from their office or other campus sites. The percentages were almost the same for the sample of researchers in this study: 40% and 60% respectively. This level of use is somewhat higher than other comparable studies of scientists. Hiltz (1984) determined that 30% of her respondents found the Electronic Information Exchange System "extremely valuable" for retrieval and searches. McClure et al. (1991) found that functions such as online database searching and remote data sources were used infrequently relative to e-mail and other computer resources.

The journal literature can be a useful mechanism for keeping current if there are publications that concentrate on the right disciplinary intersections. For the Neurophysiologist, *Neural Networks* is a key journal because it covers research on a wide range of scientific processes. This same title is also a significant source for the Psycholinguist, who has a different subject orientation. The Protein Specialist observed a change in journal publication over the past ten or fifteen years. Many new titles had appeared that were intended to fit cross-disciplinary audiences. In his research area, *Proteins* had become influential. The content of this journal would also interest biophysicists working on membranes, biologists in photosynthesis, and physicists doing drug design, among others. For network modelers in biology, psychology, physics, or physiology, *Biological Cybernetics* had become an important publication. The combined practice of browsing general multidisciplinary titles and concurrently reading selections from the more specialized cross-disciplinary journals is a way of accumulating information broadly and deeply from the literature.

Cross-disciplinary review articles are a unique type of literature that can supplant extensive information gathering by providing packages of collected, filtered, and partly synthesized information. In a 1973 study of physical scientists, Skelton found that review literature was not considered to be especially useful. Interdisciplinary researchers have a great need for sources that will help them locate relevant materials, and well-constructed reviews can assist in this problematic process and function as true boundary objects. Interdisciplinary reviews are integrative in nature, representing an intersection of multiple scientific literatures, and therefore can serve as a bridge for the researchers grounded in the fields of study covered therein. For interdisciplinary research, these compilations may be used as a concise introduction to a problem area or as a guide for catching up on new work. It follows that review writing is then a valuable contribution to interdisciplinary progress, but, as we will see, producing this type of literature is not a priority for most scientists.

Scientists do not confine their reading to the published literature, even though they do tend to emphasize it in their descriptions of their reading practices. The few researchers who talked in detail about unpublished material found it invaluable. For them refereeing, editing, and reviewing are among the best ways to find out about what is at the cutting edge of scientific activity.

> Every three or four months or so I review all this. [points to boxes of manuscripts on the floor] . . . other people's ideas, their grant proposals. And of course they are confidential so I'm not supposed to pirate any ideas from them, but you do get stimulation, what's the state of the art research being pursued nationwide. (The Structural Biologist)

Although reviewing is demanding, the current awareness benefits seem unmatched by any other strategy. Reading other people's grant proposals is probably the most timely way to keep in touch with the forefront of science.

Research Related Learning

> Every good research group strikes a good balance between learning and doing. Even a seasoned researcher must keep a good balance of learning. (Sam Smith)

Learning is a significant part of the research process, and the motive driving many information activities. Information is sometimes sought to support what is already known, and information that is collected may never result in substantial new knowledge. Nevertheless, finding, gathering, and probing for information are frequently part of a deliberate learning initiative. Learning is often the explicit goal of probing, as when researchers explore general and multidisciplinary literature to expand their knowledge base. Other import practices and combinations of activities are well suited to the pursuit of learning. Colleagues, on an individual basis, function as pointers, directing researchers to the most important and useful literature in new subject areas, helping to construct a course of effective learning. The process of collecting, evaluating, and learning from selected information is also performed by groups.

Team learning is practiced by some of the larger, more organized research groups. In order to maintain active learning environments in their laboratories, two of the team leaders developed formal group methods for gathering and filtering literature. The Neuroscience Manager described his lab members as a "roving information source"; they meet regularly to share and discuss important discoveries in the literature. The Photosynthesis Specialist organized what he called a journal club. Each student was responsible for scanning a set of journals in an area of interest to the lab members. Once a month they got together and each person presented the most interesting studies from their assigned titles. After using this technique for four years, he instituted a new procedure that added another layer of filtering and required less reading. Instead of going directly to the journals, the Center's library provided article titles and abstracts based on keywords selected by the group. The library sent about 150-300 abstracts every other week. In the new procedure, lab members chose and presented the abstracts they thought were the most interesting and posted them in the lab for everyone to read.

> At the meeting in which the student presents the abstracts, we have a little discussion about each, and we decide on one or two papers on the list that we are going to cover in more detail. The week afterwards we go through those papers in detail. Everyone is supposed to read them, and we have a discussion if there is anything new. That's how we operate our literature.

The journal club operation does more than promote current awareness; it is a learning experience, especially for the lead scientist who is in a better position than the students to judge and assimilate the work. The process begins with a systematic review of the recent literature and then the "new" is targeted for more thorough analysis that will result in adding that material to the team knowledge base. After the group has worked through the targeted papers, individual team members may proceed with another stage—footnote chasing. White (1994) refers to footnote chasing as "scholarly intelligence" (46), a practice that produces evaluated and highly conditional bibliographic information. Once an important paper is identified, a researcher will follow the channels of references through the literature. This is a standard practice for researchers and scholars in most fields, but because of the dispersion problem in interdisciplinary work, this technique at times is the best or only way of identifying important material in peripheral bodies of literature. Name-based searching is a related technique. Many researchers watch for or search out papers by the people they respect or recognize in a problem area. The Photosynthesis Specialist had experienced the limitations of name recognition as an information gathering method.

> If Joe Block published a paper and you know Joe is a bright guy, then it is going to have something interesting to say. Where if Bill Scum publishes a paper you can be pretty sure that it will be the same old stuff and you will waste your time reading it. Unfortunately, Bill Scum every now and then has a bright idea, and then no one reads it.

Joe Block has achieved a level of scientific authority that is accumulated by others when they choose to reference or build on his research. As Bourdieu (1975) asserts, scientific authority is a kind of "social capital," the value of which is reflected in reputation and prestige. There is, however, "no arbitrating authority" that can legitimate authorities: "there are no good judges, because there is no judge who is not also a party to the dispute" (25). In developing hybrid research areas, recognized authorities may not have emerged; the allocation of social capital is still under negotiation.

Using name recognition in excess, in terms of name searching and footnote chasing, has implications for building new interdisciplinary knowledge. The practice can produce highly homogeneous and biased representations of a body of literature on a topic. Consequently, Cooper (1989) warns that invisible colleges are flawed reference groups for integrative research. This situation creates a tension in literature and learning processes for interdisciplinary research: one of the most common approaches to gathering materials is one of the least likely to bring together variant backgrounds and perspectives. This seems to have been recognized at some level by the generalists who browse and probe regularly, in addition to following standard citation paths through the literature.

We have seen that colleagues in personal networks add context and meaning to new information, helping to transform it into useful knowledge. In a similar sense, reading followed by discussion appears to be one of the most valuable information routines for research related learning. This is the sequence of activities that is applied in the standard college seminar and in the Photosynthesis Specialist's journal club. This type of deliberate learning is also practiced informally among individuals. Two researchers in the sample, an experimentalist and a theoretician, combined reading and discussion in a dedicated, interactive way for an extended period of time to learn basics in biochemistry. They met regularly to discuss segments from a standard textbook. We "picked up a couple of new biochem books and met for lunch every Thursday for a year and ground our way through." A number of other researchers cited textbooks and other "basic" derivative works as good sources for learning fundamentals about an unfamiliar field.

Workshops and classes are also effective means for accomplishing the difficult task of new learning. The Psycholinguist attended a series of workshops each summer in order to "retool"—to keep up with the "complex formal systems" in linguistics. The Language Modeler, who was collaborating with a lawyer, had recently devoted a substantial amount of time to learning more about the law. Carrying out coursework is a recognized luxury, however.

> It must have been my sabbatical year, otherwise I would not have possibly had time to hang around the law school and go to an

> evidence class every day and do the readings for it. But I learned a lot doing it that way.

The quality of learning through formal classes is obviously very high. But organized teaching forums do not always fit an identified target area. Most learning is self or group sustained and aimed at certain fields that need to be covered or gaps to be filled depending on the research questions at hand. This may explain why formal opportunities sponsored by the Center to promote cross-disciplinary learning were not widely attended.

> The idea is, I am a neuroscientist and I am going to listen to this guy from computational electronics tell me in layman's language what he does and why he finds it interesting. But the program never really worked very well. I mean, the people who came to the talks were largely the people who were in the disciplines from which the faculty member came. There was not very much crossing over. (The Organizer)

The distinguished speakers drew a little better crowd but also fell short of fulfilling the objective, in the Organizer's view. While the numbers attending may not have been great, certain individuals considered these general interest colloquia an essential part of their program of learning, and in some cases the presentations that were particularly far afield were considered the most stimulating. The Neurophysiologist gave a specific example of how his research benefited from a lecture that had initially appeared to be unrelated to his work.

> So this guy came and talked about his model of swarms, swarms of ants, the dynamics of swarms of insects and how they can accomplish things. . . . I just thought it sounded interesting, and of course the guy who presented this also had the idea that this could be applicable. He didn't know where, but more broadly in a general way. When I went to the seminar I thought it very interesting, and now recently we've been able to apply a model like that to learning in the nervous system, where learning is autonomous and cooperative—where individual elements kind of search around randomly like ants, and when they do the right thing then they persist at that; they cooperate.

The researchers who made a point of going to lectures in outside fields were hoping to learn something pivotal or experience a flash of insight. To them, discovering an exciting new idea or research direction was well worth the investment in time. Even though the Center made this kind of learning convenient by hosting regular lectures, most researchers still felt they did not have time to attend. Those who did not go to the seminars understood that they were missing something valuable and wished they could fit them into their schedule. The researchers who emphasized the value of seminars and actually attended the events were interested in concept and theory development. They had made learning a clear priority in their research work.

Teaching to Learn

> Teaching is the excuse I use for reading a lot of new stuff. I only get it done by assigning it to students, and then I have to read it, because we are going to discuss it in class. I have to, if I am going to present it. I have to be able to answer all those questions. (The Language Modeler)

There was a healthy symbiosis between research and teaching for a number of scientists. They were aware of the reciprocal relationship and believed it worked to the advantage of interdisciplinary inquiry. Several of the researchers considered their teaching a natural way to "maintain a fairly broad knowledge of what is happening in other areas," while recognizing that other learning methods must be used to develop deeper knowledge. Teaching-to-learn is practiced in a number of ways, but conducting seminar type classes appears to be the most conducive to substantial research related learning. "Current topics" courses create an opportunity to explore a new area. The organization and background work that goes into designing the course is the first learning stage. The next stage comes through interaction once the course is in session, and the level of dialogue can be quite sophisticated since advanced students, postdocs, and faculty are attracted to such courses.

Short-term teaching experiences can also be useful for learning in new subject areas. Some researchers see one-time presentations as opportunities to push themselves to learn something well enough to explain it to others. The Perception Specialist motivated himself to learn by volunteering to give talks to his peers.

> One technique that has helped me understand issues that I didn't have much expertise in is putting myself in a situation where I'll maybe give a lecture on a topic. . . . So I have given lectures to psychologists on the methodology or the physics which underlie magnetic resonance imaging.

In this situation, the commitment forced him to learn the subject quickly, and then he benefited further from the perspectives offered by a group of colleagues with a similar background.

Some research projects also begin for pedagogical reasons. The Biophysicist described the evolution of one successful interdisciplinary effort that started as "a little model for how cells regulate their water content." The existing biophysics texts did not have an explanation that satisfied her teaching needs; "everything that I read seemed hokey to me." Consequently, she designed a much more detailed model to use in the class, eventually published the work, and then used the paper to teach her course. Years later she collaborated on a spinoff project with a student.

> I published this paper . . . and then I used the results as a basis for my teaching. That sort of became a little thing that was added to what I knew about the subject. And then the work in my laboratory had become more and more molecular, and [a student] came into the lab and she thought that she would like to do something at a higher level of organization, a more integrated level. She came with a mathematics rather than a physics or a chemistry background.

The Biophysicist went on to explain how pieces of a discovery came together out of the models she built for teaching, a related paper, the interests of a student from a different background, and some general reading. These components culminated into a research study that ended up having important implications for a problem in the field of medicine.

Intermediaries

Among all the various types of trajection elements, people are the most vital. They play a critical role in the transfer of information between social worlds, succeeding where literature, instruments, and methods on their own are likely to fail. In the evolution of the model described above, the student with the mathematics knowledge was the final and the necessary impetus for the new discovery. Individuals may also serve as conduits, enhancing the exchange of information by learning, filtering, analyzing, and forming intellectual connections, functioning as intermediaries, or as a kind of translator between scientific communities. From the perspective of actor network theory, defining the role of the intermediary is in itself an act of translation; further translation takes place as the intermediary serves as a spokesperson for the entities of a network (Callon, Law, and Rip 1986). An intermediary may bridge the work of two different labs, act as a carrier of knowledge between academic research and industry, or provide the link between experimentation and theory. Within the context of this study, graduate students usually performed this unique research function.

For an applied computation project, the AI Specialist trained a graduate student to work as an emissary. The student went out into the private sector for an extended period of time to live in and learn about the community, and to establish a solid connection for the future. Two of the researchers working in the team leader mode used students to gain knowledge from other academic research operations. The Structural Biologist explained that "if we don't know a certain technique we will send people to an expert's lab to learn how to do it." The Photosynthesis Specialist had arranged a kind of trading program between his and several colleagues' labs. The students cross the border to get training, and when they return they apply the new knowledge to their own work and teach others in their lab.

> Students from my lab work in Mac's lab, and Mac's students work in my lab. So my students learn molecular engineering, and some of Mac's students learn to do biophysical experiments. Through that I've got students and postdocs who do biomolecular engineering in my lab as well.

One student's internship in another lab turned out to be particularly beneficial for both sides. When he came back to the home laboratory, he had the ability to set up a new molecular engineering facility where he then proceeded to invent an entirely new method. The scientist noted that this new discovery "actually works better" than the one they adopted from the other lab and now both labs are using the new method.[9]

Frequently, information needs to be translated before it can be understood or applied. All the researchers seemed to be acutely aware of the communication difficulties across disciplinary boundaries, and a few recognized the need for intermediaries who can interpret and convey the basics about research problems and approaches. Tom was part of a project that was trying to span an extensive experimental/theoretical divide. He was responsible for the experiments, and his collaborator, a physicist, was developing the theory. They had assigned a series of graduate students to translate and mediate between them on this very ambitious project, but there were ongoing complications.

> It is not clear how to take his results and translate it into a computer file to send over to the computer to say turn these pumps on at thus and so time and run them at thus and so rate—this is what the output is supposed to be. We are now on our second physics graduate student trying to act as the lubricant to translate the two.

The student intermediary had the task of determining what could be maneuvered in the chemical world of one scientist and how it related to the symbols in the other scientist's theory. Tom seemed to be confident that with time and a lot of concentrated work, the students would succeed in the translator role, although changes in personnel had been a serious problem as the project extended over a long period of time. Before new graduate students could make a contribution, they needed to learn about the general problem area and specifics of the study. It was not clear if the translation skills developed by the intermediaries were being transferred to Tom in some way or how much momentum was lost when a student left the project.

Learning to be a productive participant in a research group takes extensive orientation and training. It requires situated learning, the process by which a person is transformed from a "newcomer" to an "old-timer" and becomes a member of a community of practice (Lave and Wenger 1991). Scientific solutions come from materials that are produced and transformed into permanent knowledge by operations that are, in essence, a craftsman's knowledge of the research process (Ravetz 1971). The intermediary craft is particularly interesting since it does not emanate directly from one community of practice. The craft is composed from the knowledge and practices of multiple scientific communities and the expertise developed in the process of negotiating between them.

Several other research groups at the Center were bridging abstract mathematical and physical worlds by applying sophisticated computational biology and physics methods. The students involved in these projects must bridge these domains as well, and the territory that spans modern computational technology and more traditional discipline-based sciences can be particularly unwieldy for them. Intermediating between the two involves blending experimental expertise with competence in current computer methodologies, each of which takes a tremendous amount of training. A student coming from a rigorous physics background will need to know or develop computer science expertise and learn the biology; those trained in computer science are likely to lack grounding in physics and biology. The demands of developing the combination of expertise can be overwhelming. Several of the scientists talked as if the expectations placed on students were unrealistic. Sam Smith felt obligated to discourage some students and counsel others along the way.

> It's very hard, and actually, I have quite a number of people who do not finish. It is very tricky, and I am a very open person in telling my students that they may want to consider not getting their Ph.D. in this. I select students early and tell a certain fraction at an early stage that they have little chance. Those who have stayed on with me actually all finished.

The Protein Specialist admitted that he was very tough on his students and that he expected them to be as diverse as him. The Movement Specialist, who had worked for the aerospace industry and government before entering academia, could not recommend an interdisciplinary research track for students who were planning to work within university structures. He thought it was unreasonable to expect researchers to follow an interdisciplinary path within the confines of academe, but he did not think that interdisciplinary training is wasted on the students who planned to work in industry. A few of his recent graduates had found positions at corporations and organizations where researchers with an interdisciplinary orientation were valued and rewarded.

THE MEANS OF EXPORT

Within the cycle of accumulation, information and other resources are imported and assimilated, rerepresented and redistributed. Information transfer and new learning takes place so that new knowledge can be created and funneled back into the cycle. The vehicles available for disseminating information to various interest groups are not always sufficient, especially when the audiences are heterogeneous. The researchers at the Center spend a good amount of their time writing about and presenting their ideas and research results. The general survey indicated that 52.8% of Center researchers spend most of their time "reporting research findings." Another 36% spend about half or a moderate amount of time on such activities. Therefore it was surprising to find that the researchers were not as forthcoming about this part of their work as they had been about their import practices. As Becher (1989) observed in his interviews with faculty members, the researchers talked more freely about their knowledge seeking activities than about how they communicate. The researchers' export processes, while not widely practiced, can be broken down into three distinct boundary crossing dissemination strategies: targeting multiple audiences, generalizing for a wide audience, and finding a fit. The researchers were clearly aware that they had multiple constituencies for their work. They described their audiences as disciplinary fusions or cross-disciplinary intersections, rarely delineating disciplinary groups. Nonetheless, most of the scientists were surprised by the degree of audience diversity represented in their individual export profile developed in Phase 1 of the study.

Multiple Audiences

All the researchers were able to identify multiple audiences for their work, while only a few actually reworked their papers for the different groups. The biophysicist, for example, had been writing both general and very specialized papers for physics, physiology, and neuroscience communities, envisioning each audience as having internal subpopulations that don't talk to each other at all. The audiences described by the Computational Neuroscientist were consistent with the notion of core and scatter. He defined his primary audience as a mixture of people at the intersection of overlapping disciplines, an "interface" between biology, engineering, and computer science, with a few smaller groups lingering on the periphery, including a select number of people in entomology and robotics. He addressed these groups separately by creating reports with a different "flavor" for each. Despite his efforts, he was convinced that many who could benefit from his research would never come in contact with it, unless they were to somehow become familiar with his name, recalling the influence of name recognition raised earlier by the Photosynthesis Specialist.

Only two other researchers, the AI Specialist and the Computational Designer, actively worked at connecting with multiple communities, with each identifying three different audiences for which they revised and presented results. The Computational Designer was skeptical about whether any kind of real "interdisciplinary community" could ever exist as long as communication channels remained separate. He found it necessary to use three different forums and three different writing styles to disseminate his work to physicists, mathematicians, and computer scientists. The AI Specialist needed to continually keep in mind who he was writing for, explaining different things in detail and highlighting various aspects related to the target group's interests. He recounted a recent experience where a conference paper he submitted on a "hot topic" in computation was rejected in the review process. In his next attempt, he rewrote the same results in a manner "tightly coupled" with the audience by using the current jargon in that specialty, and the paper was readily accepted.

The Structural Biologist broke down the audiences for his cancer studies by functional sectors: the research community, the drug industry, and people developing biomedical applications. The experimental/theoretical dichotomy is another audience consideration. According to the Molecular Network Specialist, theoreticians usually only talk to other theoreticians, and while he acknowledged that experimental research contributed to the development of his theories, he did not expect his work to influence experimentalists in return.

> Experimenters generally can't afford to be very interested in what theoreticians say; . . . experimenters will give a congenial, perhaps amused, or bemused hearing to theoreticians just to find out some new slant perhaps, which they would incorporate into their own thinking about how to do experiments. Rarely would I get experimenters reading my work.

The Neurophysiologist doing biophysical modeling was disillusioned by his inability to get his work accepted by the experimental neurophysiology journals, the only publications that would reach his primary audience. "They occasionally publish a model, but not very often, and if they do it has to be a very traditional type of model." He recounted that when he was a student, even his doctoral advisor questioned his penchant for modeling, warning that "we do our experiments and *then* we know the data."

General Audiences

In some research areas experimentation and theory do seem to be melding. Reflecting on the impact of this type of boundary crossing on progress in bioenergetics, Neil explained that computational development has facilitated a "natural evolution of experimental and theoretical methodologies," bringing together what had been "separate for traditional reasons." From his point of view, improved communication was the prime factor in the successful synthesis of theory and experiment developed in different scientific domains. His communication approach de-emphasized theoretical formulas. "I use simple physical chemistry, and some simple kinetics. But I would never show an equation with more than a couple of terms."

Other generalists stressed the importance of generalizing their papers. A common tactic for writing for a "general yield" was to include a preface of introductory material that could get a wide range of scientists oriented to the work. After the introduction, however, the text would quickly become much more detailed and specialized. The Neurophysiologist and the Movement Specialist reported limited success with this technique in practice, since journals adhere to fairly rigid parameters that are aligned with what is perceived to be the dedicated audience.[10] The researchers reported more success presenting broadly based reports at conferences or symposia. The problem-centered conferences that these scientists attend have diverse participants that require more generalized discussion. Presentations and other face-to-face group formats provide interaction with the audience—a process of communication, rather than a static reporting mechanism. Audience feedback is useful for gauging the level of a talk as it progresses, but it is not uncommon for a researcher to underestimate the foundational material that needs to be incorporated:

> I would have to explain the stuff that I had not thought about having to explain. Even though I started with the assumption that they didn't know anything about this theory, there was all of the stuff that's just background. (The Language Modeler)

Cross-disciplinary audience dynamics can transform a research report into a teaching session of disciplinary fundamentals.

The Neurophysiologist was concerned about the increasingly technical nature of scientific literature.

> I have always found that it is okay to be general. . . . The experts don't mind, by and large. Of course, some do. Some will be critical in their own way. But I don't think that it is appropriate in this day and age for interdisciplinary work. I think science is shrinking. The number of experts in any one field is going to be smaller, so you are going to have to be more general if you are going to want to talk to your colleagues.

He favored the use of analogies in his talks to multidisciplinary groups. For example, by explaining how the action of the vestibular system is similar to a shock absorber, he would give a group a conceptual grasp of his topic before he moved into depth on more specific functions. The Humanist also spoke of a crisis in modern science exemplified by specialized journal articles that can only be understood by those who write them, putting the public at the mercy of the equipment and structure of big science. The technical nature of the scientific literature, according to Latour's (1987) analysis, is a significant barrier made up of layers of experts and their symbolic authority. As authors bring in more sources, or "many well-equipped men's words" and their many associations, it builds up a disproportionate amount of social linkages and the literature becomes more technical and inaccessible.

Unlike the Neurophysiologist, some researchers were not compelled to work at reaching diverse audiences. To them, rewriting is not a worthwhile use of time.

> It's not in the writing, but the very nature of the work that is appealing to them, so if it is the right thing, appropriate thing for them, then it is not even important that they be written for. I think both sides are smart enough to get what they want out of it. Just as we go and look up their papers and just get the 10% we want.

In this scenario, import continues to be the primary thrust in the cross-disciplinary information transfer process, while active export is not considered necessary. The responsibility for finding and interpreting the information lies with the reader; receivers must take on the translation work.

Review literature, as noted in the last section, serves an important integrative role in interdisciplinary accumulation. Researchers seek out reviews for their own consumption, and a few create these works for export to others. Reviews may come about as a byproduct of a research project, an important outcome of a difficult information gathering effort that can be shared with newcomers to a community of practice. Tom was involved in writing a review of this type.

> It's not just out of altruism, believe me. It's that whenever somebody comes into this group, if they don't have all the information where they can get their hands on it, then generation after generation of students are going to be hitting their heads against the same problem.

As with other emerging hybrid areas, the complex systems literature base was young and undeveloped. In the past, Tom could have confidently stated: "if you start at the beginning and go to the end of this body of material, you're ready to roll." Now there is no standard reading to recommend to new lab members or graduate students, just the compilations constructed through the course of a project.

Apparently, boundary crossing review writing works well in teams. Tom had involved a collaborator from another university and a research assistant in writing a recent review article. The Neuroscience Manager periodically holds a seminar in his lab to "amass information on a particular topic and coalesce it into a collectively produced review piece." Several other researchers mentioned the importance of review literature, but according to Tom, there are not many incentives for taking on this type of project at a research university. "There is some feeling that anybody can write a review, but it takes a real first-rate scientist to do experiments."

Finding a Fit

The number of interdisciplinary and problem-centered journals is increasing, making it easier for researchers in some areas to find an outlet for their work. The Protein Specialist, who received his Ph.D. in 1982, pointed out that when he was a student there were no periodicals that bridged computer science and chemistry; now there are at least four. Journals like *Proteins* and *Protein Science* are not merely adding computation to their focus, they also bring together physics, biology, and chemistry. He praised the fact that certain established disciplinary journals had adopted a more multidisciplinary scope, such as the *Journal of the American Chemical Society*, which now accepts work from physics and biology.

Still, certain researchers have difficulty placing their papers. As mentioned earlier, the Neurophysiologist had not been successful getting his modeling work accepted in the journals that represent his primary audience. The Human Factors Psychologist, who writes for the "human factors people," believed that his research is perceived to be too "far out" by most psychologists, and, since he challenges some established ideas in the human factors community, his work had not been welcomed there either. He had recently presented a paper to a new audience at a virtual environment conference and received an unusually favorable reception.

> There were all of these people from a whole lot of different sciences there. I believe that they really liked my paper. I felt really good about it. There was a lot of discussion and it really worked well.

The conference attendees had a keen interest in the topic, and they readily understood the application of the research. Since this audience did not consist exclusively of people from psychology or human factors communities, the methods and results of the project were not being assessed by criteria developed within those established fields of study. That is, an audience that represents a discipline or core specialization would be more inclined to base their evaluation on current disciplinary standards. A group of boundary crossing researchers may not have the background to critique the work from that perspective, and their aims may, in fact, be quite different. They are congregating based on a common problem area, not a common base of knowledge or disciplinary culture. Moreover, their goal at such a gathering may be to learn about new approaches and complementary initiatives, and to identify other researchers who can help them navigate among the diverse subject areas that are relevant to the problem.

Instead of finding a match, some researchers struggle to make a fit between their research and the existing dissemination venues. The most common strategy is to customize research reports around the formats and focuses accepted by the recognized journals. There is, however, a risk of crossing the line between "making a fit" and "forcing a fit." The Computational Designer gave a detailed account of the labor involved in getting an applied physics paper accepted for publication. In his opinion, the available journal options were disappointing. They were either too formal or too applied; his particular audience was situated somewhere in the middle. He was still working through a difficult and frustrating review and revision process[11] with one of the more formal publications in which two years had been spent "in a haggle over how the final version of the paper should look." He was confident that he would succeed in getting it published, but:

> It takes forever because they say, well this is not how we usually say things, and you have to say it differently. Then you rewrite it for them, and they are still not happy. It is very hard. It takes very hard work to preserve the integrity of the paper, because if you go too far their way the paper is very easily hurt.

In this case there seems to be a misalignment between the reviewers and the target audience, if, as the Computational Designer predicted, "the community of people who have to use it are going to say: this is a very difficult paper, we can't read this paper." The journal was clearly not the right fit for the work, but it was the best available option.

EXCHANGES

As discussed early in this chapter, many information practices support both import and export or the exchange of information. This is especially true of conferences, e mail correspondence, and intermediary activities. The notion of exchange is the essence of interdisciplinarity, in that the dialectic process can be considered the means of interdisciplinary inquiry. It is a process of discourse "between two or more speakers who express two or more positions or opinions," where the speakers represent disciplines and the differences are ultimately resolved to form a synthesis (Klein 1990, 194). Neil was convinced that significant progress could be made in the discourse as individuals learn to communicate across boundaries.

> Over the past six to eight years the very physical types have gotten better at giving pictorial insights, or you know, qualitative insights. I mean it has really helped, and it has forced them to interpret quantum physics, not just in lay terms, but in real mechanistic terms. . . . So forcing them to talk at the mechanistic insight level that a biologist is comfortable with actually forces them to examine what they do. At the same time, the biologists who ten years ago would have fainted at the thought of having to understand anything quantum mechanical are forcing themselves to learn at least the vocabulary of quantum theory, and to come to terms with things which maybe they can't or never will understand at the deep level, but for which they can begin to appreciate the framework upon which mechanistic questions can be based.

The exchange between biology and physics fuses the disciplines by enlarging the scope of understanding and offering more opportunity for insights. Neil attributed significant advancements over the past decade to physicists and biologists beginning to explain things to each other.

There was a prevailing expectation among the researchers that the rate of exchange between scientists would increase rather dramatically in the near future through the use of networked communication systems. The Biophysicist, who had recently written an expository paper on interdisciplinary science, predicted that "the most meaningful continual interactions among scientists are going to be in cyberspace." She suggested that at present these interdisciplinary electronic networks seem to exist "as implicit communities, consisting only as persistent patterns of circulating e mail about research news from laboratories working on closely related problems." Reports in the early 1980s predicted a similar scenario, suggesting that cross fertilization would stimulate inquiry as researchers began to explore the applications of electronic information networks (Turoff and Hiltz 1980), and that new electronic invisible colleges would give rise to a liberating collective intelligence (Tracz 1980).

Glimmers of new electronic communities and their reliance on computer supported cooperative work were evident at the time of this study. Researchers did not talk explicitly about sharing information electronically, but they were enthusiastic about doing cooperative work over e-mail. Clearly, in this context, e-mail was the primary tool for information exchange. E-mail facilitated long distance cooperative research for a number of the scientists. The Psycholinguist described a current collaborative arrangement with neuropsychologists in another state. "They had some really good ideas, so we've been working mostly on e-mail. E-mail made all this possible." The Language Modeler talked enthusiastically about the paper she coauthored with a lawyer, exclaiming: "I still can't tell you what he looks like." E-mail is also used for queries and consultation but is not always the format of choice for such interactions. A number of scientists still prefer to be able to "pin someone down and pick his brain" in a face-to-face exchange.

As was true with the other information mechanisms discussed in this chapter, electronic import was emphasized more than export by the researchers. Few participated in forums designed to promote information exchange, such as bulletin boards, listservs, and electronic conferences. The general survey of information use at the Center indicated that 35% of the members never used electronic news groups and bulletin boards; only 23% used them regularly, a few times per week. When they did connect to bulletin boards and listservs, the researchers tended to lurk rather than contribute. The electronic forums were perceived as inefficient sources of information with low "signal to noise ratio" and poor quality content—"junk mail," "gossip," and "chatter."

Several scientists expressed a different reason for not using or learning more about electronic communication options. They simply felt that they already spent too much time at the computer and had made a conscious effort to find ways of working that did not involve sitting at a terminal. Going to the library, browsing a journal that comes in the mail, or walking to a colleague's office to ask a question were welcome activities. For them, learning new computer techniques was a "time pit," with the cost of learning the new technology greater than the perceived benefits. The computing systems staff member I interviewed had been involved in a recent campaign to upgrade the networking capabilities for the researchers at the Center. He observed that the graduate students gave him more feedback on their information needs and uses than the faculty members did. His experience was that the researchers were not aspiring for greater networking capabilities; they seemed to be satisfied with the current system simply because "it works."

The information practices and processes of interdisciplinary researchers are varied and complex. The scientists utilize formal and informal information channels, apply individual and group approaches, and take advantage of written, oral, and to a lesser degree, electronic formats. In their attempts to "go at it from all angles," they seek, gather and probe for information, and some target, generalize, and customize their work for various audiences. Specific patterns can be discerned within the research modes. For instance, in terms of reading, generalists tend to be browsers while collaborators are more focused readers.

Moving beyond the individual to working communities of scholars, it is more accurate to emphasize the problem-centered and multimodal nature of the research enterprise. Interdisciplinarians and their communities of practice employ an aggregation of processes to tackle the problem at hand. Accordingly, boundary crossing inquiry requires a repertoire of import and export practices. Concerted communication and exchange practices appear to be the most vital to making research progress. In particular, the multidisciplinary composition of personal networks and problem-centered conferences allows interpersonal exchange across disciplines within meaningful parameters. The specific strategies emphasized by the researchers also tended to be exchange-based. Tom and the Molecular Network Specialist learned best through discussion, as did the journal club members. The physical exchange of students between labs led to the successful development of a new biomolecular engineering method. In the concluding chapter, I will apply what we have learned about boundary crossing accumulation practices to suggestions for the organization of information and the development of interdisciplinary information services. At this juncture, we will turn away from specific work strategies to examine the structures and climate of interdisciplinary research work at the Center.

[1] This researcher's engineering orientation may account for his heavy reliance on colleagues. While pure and applied scientists working in universities, government, and industry prefer informal information (Rosenberg 1967), Allen's (1966) information flow studies found that technologists rely more on oral sources than scientists do.

[2] Research in management has shown that when there is high uncertainty in a situation, decision-makers prefer oral over written communication (O'Reilly 1982).

[3] Recall that to the humanist the convergence of science was the most significant function of the Center.

[4] See Bernier and Yerkey (1979) for a discussion of information gatekeepers as filters and Neill (1989) on the role of the information analyst as quality filter.

[5] Sproull and Kiesler (1991) document similar advantages of electronic communication for coordination of group work in organizations.

[6] According to Garvey and Griffith (1964) those who present at conferences are often in the upper echelon of science, and they are likely to be more productive and publish more than researchers who do not present their research.

[7] According to Oseman (1989) only about 1/5 of all conferences are meetings of large societies; most are organized around specific problems or topics.

[8] In their survey on the impact of electronic networks on scholarly communication, McClure et al. (1991) also found that researchers commented most about the ability of networks to enhance interaction between colleagues.

[9] In his study of the U.S. steel industry, Eric von Hippel (1988) links innovation to informal "know-how" trading that takes place between companies. He demonstrates that even rival firms exchange specialized knowledge within networks of engineers with common research interests.

[10] Ethnographic studies underscore the forces, other than the author and the interpretive community, that affect writing and reading. See, for example, Clifford and Marcus (1986) and Boyarin (1992).

[11] See Myers (1993) for a detailed account of how two biology papers were reconstructed through the peer review process.

Chapter 3.

CONDITIONS OF BOUNDARY CROSSING RESEARCH

The researchers' information activities do not take place in isolation. Their ability to import and export across disciplinary boundaries and conduct integrative research projects is moderated by the organizational and cultural conditions of their work environment. What forces affect interdisciplinary information practices and processes, and what are the conditions of work that lead to successful boundary crossing research? The researchers' everyday work practices, affinities, affiliations, and interactions direct attention to certain problem areas and in time work to change the shape of disciplines. Global and local organizational structures also influence what problems are addressed and how individuals investigate these problems. In the scientists' work worlds there are centrifugal forces that help to move people, things, and ideas outward into other domains, and centripetal forces that work to hold elements together in established frameworks. The Center's research environment has been deliberately designed to facilitate boundary crossing research activities, and many scientists thrive in this climate. There are others who, despite the Center's amenities, spend little time there. Their cases highlight features and predilections of interdisciplinary research that are not dependent on the Center's resources. In this chapter, we will revisit a number of concepts introduced in Chapter 1 to consider the structural and cultural forces, both within and outside the Center, that govern how interdisciplinary research work gets done.

DISCIPLINARY OBSOLESCENCE

The researchers expressed concern about the configuration of scientific fields in relation to their individual research and the overall progress of science. They did not consider current academic structures an accurate reflection of present or future research opportunities. According to Sam, the academic world is conservative and represents an outmoded era of science that only rewarded disciplinary specialization. He believes that the discipline-bound problems have largely been solved[1] but that the disciplines did play an important role in getting to the current state of knowledge.

> We learned to concentrate and focus well, and I hope the lessons we learned from that we will never forget, and not just have people who are shallow and know everything and nothing well. . . . And now the problems that are not addressed by them have to be addressed so they are not falling through the cracks. That is why there are other combinations of knowledge.

Sam described the disciplines as "lampposts" that shed light on some parts of the intellectual panorama; the dark spots between the illuminated areas are where the opportunities lie now. Scientists are most likely to look for answers in the lighted areas where they can "see better." He suggested that they have a tendency to concentrate on problems they can do, not necessarily the problems that are most crucial.

> Rarely do scientists proceed as one would expect: realize a problem, identify the logically best path to the solution, and then try to follow it. Rather, they just do what they can do, hoping the result is of some value. Naturally, this approach is a key pitfall of established disciplines . . . often they force their approach onto outside fields which possibly could be served better with less elegant or complex approaches which, however, are below the dignity of the established disciplines.

The lamplight metaphor is reminiscent of Campbell's (1969) conceptualization of the disciplines as isolated islands of clustered specialties, with the space between the islands analogous to Sam's dark spots. Campbell proposed a fish-scale model of omniscience as a more effective alternative for the organization of knowledge. With overlapping, continuous layers and no rigid boundaries, this structure can embody depth and has the potential for unlimited breadth. In practice, such a framework would be an interdisciplinary ideal; it would allow for unhindered cross-disciplinary communication and promote the development of new combinations of competencies.

Sam was not alone in questioning the future viability of existing academic disciplines. The computer scientists were particularly candid about the limitations of the current conceptualization of their field. The Computational Designer believed that computer science could become obsolete.

> I think that the contours are becoming visible—if computer science doesn't give up its autonomy it is going to lose its prestige. . . . There is no point in doing these things in their own right.

He objected to the discipline's formalistic tendencies, claiming that computer scientists must integrate with other areas to create useful applications. The AI Specialist expressed similar views. He argued that applied computational work should evolve in concert with problems and users' needs, instead of within the existing disciplinary confines.

> It's really embarrassing—[computer science] people who hear about the needs of scientists, or anyone else for that matter, and then try to shoehorn them into some solution that has been done in the past, which is particularly inappropriate. . . . It is just completely the wrong way to sell it to them, and from the technical point—to solve the problem. Just any other way I can think of, it's the wrong way to do it. Yet, since it is the way it was done in the past, they are coming in and doing it now and it's a terrible, terrible mistake.

In fact, as he explained it, the problem is not new: computer science moves so fast that every five or ten years the community must "reinvent itself" and "collectively haul itself back onto track and into relevancy."

Biology, on the other hand, appears to have successfully converged with related sciences and to have begun cultivating new approaches and addressing new problems. Biologists have succeeded in creating what the Biophysicist called "new physics."

> I guess when you are elucidating the physical basis of phenomena that do not exist elsewhere, and really haven't been described in any other context, to me that's new physics. But it doesn't really matter whether you call it new physics or not—it's new knowledge.

This new field is focused on the "complexity of life." The biologists and physicists have watched their fields fuse over the years and mature to the point where the scientists no longer need to aggressively explore and gather knowledge from the other disciplines.

> I'm finding really that the multidisciplinary lines are coming into me, so that I have less far to reach. If I was doing this 25 or 30 years ago, then I think one would have a very long reach. . . . The environment in many ways is getting smaller, and I suppose one is reminded of the

> view of the world as being a global village through the advocacy of computational networks. (Neil Nash)

The changes in biophysics journals, such as the introduction of new problem-centered titles and a broadening of scope in established disciplinary journals, are further evidence that this hybrid field has reached a state of maturity.

INSTITUTIONALIZED INTERDISCIPLINARITY

Manifestations of the hybridization process, as conceptualized by Dogan and Pahre (1990), take various shapes in the university-based scientific enterprise. Within universities specialized units of knowledge come together and are formalized in cross-departmental programs, interdisciplinary colleges, and separate research organizations like the Center. The Center houses project teams and individuals whose hybrid research interests do not fit into a single academic department or discipline. From Sam's perspective, the Center's structure is compatible with the real questions and opportunities in contemporary science, providing "a cradle" or sheltering environment that can give rise to new scientific fields. He believes that institutions that support problem areas rather than disciplines are essential to the process of disciplinary evolution in that scientific disciplines begin with problems—"technical problems, medical problems, and intellectual problems"—and new knowledge is free to develop when research work is mobilized around problems.

Despite its problem orientation, in some respects the Center exemplifies Lenoir's (1993) idea of a discipline building institution. It affects change in the intellectual landscape through the creation of an "institutional niche," a place where new routines can be supported long enough for distinctive types of work to emerge. Following Latour's approach to studying science, it is important to examine the network of actors, practices, and resources in relation to each other and not place undue emphasis on the supporting institutions. Nevertheless, the nodes of a network are not all equal, and their differences are central to understanding the dynamics of the network as a whole. In interdisciplinary work the aims, politics, and outcomes of the research process can diverge from those espoused by the associated research institutions. Interdisciplinary researchers are affiliated with multiple discipline- and hybrid-building agencies, and they have different relationships with each. The complex organizational ties interdisciplinarians develop are part of how they do their science, and these relationships are built and sustained through strategic information and communication processes.

Disciplinary programs are institutionally oriented, according to Lenoir. They are concerned with "establishing service roles, facilitating links with other disciplines," and enabling transmission of techniques and tools (79), goals that are quite similar to those of the Center's interdisciplinary programs. Research programs have a different purpose. They are directed at controlling and defining a scientific field. They strive to "dominate the cycles of credit and available resources for extending and legitimating products of their research" (Lenoir 1993, 79). Clearly, the disciplinary programs of organizations and research programs of individuals are deeply intertwined. Institutional disciplinary programs, or interdisciplinary programs as in the case of the Center, enable the progress of research programs; in return the steps researchers take to control and legitimatize their emerging research domains sustain institutions and their disciplines. Interdisciplinary research programs are situated across more than one discipline building institution and can therefore marshal people, facilities, money, and other resources from multiple sites, that is, if the researchers involved in the programs can maneuver between the work cultures of the various institutions.

Strauss (1978) and Clarke and Fujimura (1992) have addressed the importance of "the site" in their studies of situationally constructed work. The Center is an especially interesting and complex site that reconfigures people and resources within the larger university structure to support a special kind of work arrangement. This administrative action formalizes the intersection of disciplinary social worlds. It also provides a base for interaction between differing "lines of work"; as people cross lines or participate in two or more lines of work, the traditional organization of work is altered (Gerson 1983). Relating this to Lenoir's distinction between the research process and research institutions, the Center provides a foundation for new routines and distinctive patterns of work to take hold, as well as a framework for disciplinary communities to intersect, develop roles, and build connections. Different ways of working are coordinated through projects and research groups that capitalize on institutional resources to advance specific interdisciplinary research programs. Lines of work converge in the meeting rooms and laboratories, where biologists, physicists, chemists, computer scientists, and members of other communities all "rub shoulders."

The discipline building academic departments enforce an administrative and educational framework but are less effective as interdisciplinary research sites; they lack the intellectual intersections and the physical interactions required by interdisciplinary scientists. Researchers who previously worked outside the bounds of the administrative frameworks to communicate, congregate, and work across disciplinary boundaries were selected to populate the Center, giving structure to what had been an informal and loosely organized system of alternative research groups. The Center established a place—a common ground where, for instance, a neuroscientist, an entomologist, and a vision specialist could come together to design and build an insectlike robot leg. The institution exerts a strong centripetal force, drawing together progressive research initiatives into a single physical site.

Organizationally, the Center is divided into research groups that serve as subsites for the researchers. The titles given to these units represent the general topic covered by the group. Interestingly, while the designated administrative titles of the research groups provide a reasonable rubric for the researchers' local reference group at the Center, they do not adequately depict the scientists' greater intellectual community that includes additional research affiliations across campus and connections with scientists located elsewhere. Problem areas have greater inertia than the organizationally designated subsites. The researchers talk about their work in terms of the people, methods, vocabularies, and information within the larger network that forms around the problems they address, rather than in terms of their local designated research group. As the figurative common ground, the problem areas—oscillating reactions, photosynthesis, and membranes—are more fluid and responsive to interactions and progress among fields.

Provisional or ad hoc institution-based sites influence research work as well. We have seen that vital activities, such as the exchange of information and the generation of ideas, are expedited within the conference framework. Recall, for example, the Human Factors Psychologist who benefited from an important dialogue with a new audience at a virtual environment conference. His contact with computer scientists moved him further away from the psychology research community into a more diverse group of scientists centered on human-machine interface problems. The conference itself was an effective temporary structure for the intersection of social worlds and interaction between their lines of work.

SHADOW STRUCTURES

The researchers may consider disciplines outmoded and academic departments misaligned with their research activities, but the Center did not become the alternative structure of choice for all of its initiates. Less formal cross-disciplinary sites, or "shadow structures,"[2] became the primary working structure for some researchers, especially those involved in cross-disciplinary academic programs rooted in educational initiatives. The "biophysics faculty" is an example of a vital shadow structure of loosely organized scientists from biology, chemistry, biochemistry, physics, agriculture, and the medical school. A research branch of this group was awarded a position at the Center when it opened, but the shadow structure remained the dominant reference group for some of the members. A similar shadow structure existed in neuroscience, where faculty from the departments of psychology, physiology, and cell and structural biology contributed to an interdisciplinary program of study for graduate students. These educational programs helped advance research programs located at the Center by generating a student work force with an interdisciplinary orientation.

> Because of the neuroscience graduate program, the students that come into my lab are sort of self-selected for being interested in interdisciplinary research and combining the biology and the modeling. (The Computational Neuroscientist)

Reciprocity between academic and research programs is not a new phenomenon, but, as Klein (1996) asserts, the balance between formal and shadow structures is changing. As knowledge becomes more heterogeneous and interdisciplinary, and as the skew between research problems and the organization of academia increases, alternative platforms for both research and education become more essential.

Shadow structures are a functional and a cultural development, and naive attempts to create them by merging separate academic programs are likely to fail. The Movement Specialist had been involved in cross-departmental initiatives between bioengineering and engineering psychology on more than one campus. In his experience, actual integration between the areas was limited to putting engineering and psychology students together into a single classroom. These efforts only achieved one important end—they made it necessary for instructors to shed their disciplinary assumptions. The programs did not, however, stimulate interaction between the social circles of research faculty.

> They are still two different cultures. There are people who just have their offices and go to the brown bag lunches in psychology and then there are the other people with offices in mechanical engineering who just go to the industrial engineering sack lunches. There are a few that cross over, but not many.

The successful interdisciplinary shadow structures discussed by the researchers tended to develop around research interests, not educational goals. When the Center was first being organized, a number of these self-organizing research communities were admitted to the new institute. It was a rare opportunity for the groups to centralize their efforts and gain institutional support. The cognitive science group proved to be the type of research enterprise that could benefit most from the Center. It was a loose intellectual community with no established academic program, and it ended up making substantial progress after it established a foundation at the Center. The Organizer noted, "the cognitive scientists are immeasurably stronger on this campus now than 6-8 years ago because of the Center." The Language Modeler called the Center "a real godsend for cognitive science. Cognitive science couldn't do anything without it because we wouldn't be in a single place." For the linguists, psychologists, educational psychologists, and anthropologists that made up the group, the physical niche was a critical factor in getting organized and becoming productive. In this case, the Center was clearly a stabilizing centripetal force that attracted and held together research efforts and resources and that enhanced both the scientists' programs of research and the hybrid building ability of the institution. One important outcome of this group's newfound stability was the power to recruit promising cognitive scientists to the university. The abundant resources and prestige of the Center attracted talented researchers to the campus, further advancing the research program and the institution.

The complex systems group also began as a shadow structure on campus, but its move to the Center did not significantly change the dynamics of the community. The problem focus of the group was central to the Center's overall mission to study "living and non-living systems of varying complexity," and, like the cognitive science group, their membership was diverse, with representatives from physics, engineering, math, and chemistry. Unlike the cognitive scientists, however, this informal group was a solid, functioning federation of researchers before moving to the Center, and it did not flourish in response to the influx of facilities and resources. In fact, the group maintained a level of independence from the Center, as displayed in Tom Tate's strong bond with his colleagues as a base for his research activities: "Where the complex systems research group goes, that is where I am going because those are the people I work with on a regular basis." Tom also noted that the group remained relatively isolated from the other groups housed at the Center. "There is some cross-fertilization, but it is mainly within CSR [complex systems research] rather than broadly across the institution." One member suggested that the complex systems scientists did not really need the Center as much as the less established groups and that it would not hurt their research efforts to go back to their original distributed arrangement.

The Center's initial selection process did, in a sense, accumulate the interdisciplinary shadow structures that already existed on campus. With their strong cross-disciplinary intellectual alliances and research interests that fit the institute's mission, the cognitive science group and the complex systems group were both prime candidates for induction. Looking back, we can see that the new environment energized the languishing cognitive science group. In contrast, the original complex systems arrangement had been highly functional. That group was better positioned to launch into new interdisciplinary projects from its new base at the Center, but it did not make obvious advances in response to its new location. The soundness of the group, however, certainly contributed to the overall institutional program of the Center.

CULTURAL FORCES

Consolidating interdisciplinary research initiatives and fostering their progress is not strictly a structural problem. Separate research cultures resist integration in the same way as the cross-disciplinary educational programs that the Movement Specialist experienced. Inasmuch as the scientists and their efforts were brought together by the Center, disciplinary norms and traditions worked as a counter force against coalescence. A member of the cognitive science group described what it is like to pursue interdisciplinary work at a university that is highly respected in conventional disciplinary areas of study and research. The departments act as "black holes."

> It is a gravitational force that can overpower you. The departments are entrenched in what they are good at, what they do well, and it works to suck you down into these big holes. The force pulls you back into working within the areas that have been done so well for so long.

As White (1987) suggests, perhaps fields of science should be thought of as "force fields," in order to accurately capture the "aggressiveness of modern academic life." Members of the Center were not completely buffered from pressures to follow conventional research tracts and methodological approaches.

The cultures of intellectual communities are formative, overriding the local administrative structures of research. In their exploration of faculty work and technology transfer, Dooris and Fairweather (1994) found that while organizational structures influence behavior, the values and norms of academic disciplines matter most to faculty at research universities. Members of disciplines develop common ways of seeing and interacting with the world (Becher 1989); they become acculturated. Our understanding of science as culture has advanced considerably since sociologists began to study science as a community-based activity (Merton 1973), with recent studies examining the shared value systems, familiar ways of working, and disciplinary rhetoric within scientific cultures. The values and practices at the foundation of coherent and vital intellectual communities can act as deterrents to boundary crossing research. Interdisciplinary researchers need to gain entrance to outside communities, and this requires a sophisticated understanding of the values, methods, and languages of their cultures. Pinch (1990) reminds us, however, that disciplinary knowledge and culture are not fixed but are constantly redefined through the activity of science. Thus, as researchers overcome cultural barriers by communicating and working with other communities, they are revising the landscape of science.

All of the researchers studied had an appreciation for the cultural dimension of interdisciplinary research, stemming from their experiences becoming members of multiple scientific communities. Noting the phenomenon of mobility between research areas, Becher (1990) suggests that "promiscuity" is a good thing in respect to group memberships in academia. Simultaneous affiliations can be an "effective means of transmitting ideas and findings from one cognate field to another, and of bringing to light key similarities in thinking between two cognate specialisms" (344). However, the notion of promiscuity suggests more frequent and indiscriminate participation in outside intellectual communities than that practiced by the researchers at the Center. Gaining entrance into new groups is one of the most challenging requirements of interdisciplinary work, and promiscuity would be beyond the means of most academic researchers. Breaking into a new research area is demanding work, and acceptance into a new intellectual community is a significant individual achievement. It took Tom more than a decade to become part of one circle of researchers.

> I've been at all the conferences that have ever been held, starting in 1982. . . . But I didn't really feel I was part of the inner group until this most recent one this year. Prior to that there was always some feeling that people who know what they are doing are over there. I'm over here, and I'm not part of them.

Like any kind of social gathering, it can be very uncomfortable to be an outsider at these critical events. Sam gave a pained account of his culture-crossing experiences at conferences.

> When you are changing disciplines and evolving into some new areas, it creates difficulties for you psychologically and socially. I'm forty-seven years old. I go to a meeting for the first time because I realize that it is important, and here comes this forty-seven year old guy that nobody has seen—obviously he is a total loser who didn't get anything done yet. That's how they treat you. You know you must almost have a masochistic tendency to like this. . . . I absolutely hate it with every molecule in my body, but that's the way it is. So you have to accept to learn, accept in a sense that you are really a lowly student.

It takes extensive time, effort, stamina, and humility to become a recognized member of a new scientific community. The process is akin to developing a new identity. Building a professional identity is a basic stage of development for newcomers to a "community of practice"—learning and identity building are inseparable processes (Lave and Wenger 1991). Adding to the complications, there are few concrete incentives in the academic environment for a researcher to shift their status from respected expert to unknown novice, beyond the opportunities provided by interacting with a new set of colleagues. Learning how to be a different kind of scientist engenders more potential risks than rewards.

There were a couple of researchers who found it as difficult to be an outsider at the local level as the national or international level. Two in particular had yet to find a comfortable niche for their research at the university. They lingered on the fringes of existing structures, detached from their departmental colleagues as well as the alternative groups at the Center, and with no other sustaining shadow structure in place. The Computational Designer is an illustrative case. He stated that he rarely talked to other people in his department about research because no one there understands what he does. He had a collegial relationship with faculty in other units, although an official attempt to transfer to another department a few years back had failed. The Movement Specialist was similarly isolated in his home department. After more than five years, he was still unsure of how to be an effective member of the faculty in his department.

> It is very difficult to do if you are spread over a number of different communities. People don't see you a lot. They don't recognize the conferences you go to or the journals you publish in. . . . I think there is a lot of mistrust of people who come from the outside. . . . Does this person value what I value? I mean I think this is a legitimate concern. It makes it difficult to integrate into a place.

He believed that interdisciplinary scientists working within the confines of academia are "doomed to failure" or at least discontent. To the Computational Designer the Center was merely extra office space—"real estate." The Movement Specialist gave the institution more credit, admitting that his affiliation had given him a "certain measure of credibility" in the larger world of science. We will see in the next chapter how important this credibility can be for supporting research that is not recognized within established disciplines.

These two particular cases are outliers that could possibly be explained by personal or political situations not revealed in the interviews. While these two researchers had yet to meld into the Center environment, others who felt displaced in their academic departments found the Center to be an essential oasis. Generally, the Center fulfilled important voids for most individuals, albeit in different ways for different scientists and groups, and not necessarily in the ways that had been anticipated. The planners of the Center sought groups that had potential for reaching out and interacting with other groups. The cognitive scientists did not fit this profile. Instead, they were a loosely coordinated, high-potential community in a fairly well established hybrid field, and they ended up to be the best match for the Center's properties.

CROSS-CULTURAL COMMUNICATION

White (1987) states that scholarly and scientific articles and books are "affirmations of the specialized discourse in which they are written, and beyond that, of the specialized community to which they are addressed." This is how communities of discourse are "constructed and maintained" (8). We have seen that communities are also reinforced through personal networks and conference-based communication. The complication for interdisciplinary communication, as articulated by the Humanist and by Neil in his Deep Space Nine analogies, is not the mode of interchange but the language. The various communities use the same communication channels, but they speak different languages. The cultural differences between groups are accentuated by their inability to understand each other's dialects and jargon. Researchers experience language barriers in all types of boundary crossing activities, including information gathering, collaboration, and informal discussion. Speaking specifically about scientific literature, the Psycholinguist called this a "comprehensibility" problem. He explained that a journal article with high comprehensibility could be widely cited in other disciplines but that low comprehensibility is the norm, with most papers laden with overly complex or technical content that is meaningless outside a select community.

Words can have different meanings for different communities or, in some cases, one meaning for one group and many meanings for another. As a linguist, the Language Modeler was particularly intrigued by this phenomenon, offering numerous accounts of cross-cultural communication difficulties. He described one particular experience in detail, an encounter with an article in a newspaper on gene splicing. Although it had been written for the general public, the language was imprecise, making it impossible for him to understand the content. He recruited a biochemist to help him translate.

> We just were not communicating. He was thinking that I was really stupid, and I was beginning to think that I was really stupid, and just, boy this stuff is just so complicated. I couldn't possibly understand what this article was even about. If I couldn't understand it, how would other people who weren't biochemists? . . . And the bottom line, after 45 minutes of talk he agreed that yes, they were using gene in five different ways . . . but it was real work to come to that. I was sweating bullets by the end of that.

The different contexts for the word "gene" were familiar, and therefore assumed by the biochemist. The Language Modeler's lack of background in the area limited his ability to understand the context and the applications of the term.

In the past, the Movement Specialist, who switched from physiology and engineering as an undergraduate to psychology during his graduate work, had similar problems on a regular basis. He remembered not understanding the context surrounding words in his newly chosen field and the discomfort of being someone who "talked about and thought about things" in a different way.

> The baggage that came along with the words for a psychologist wasn't necessarily the baggage that I was aware of or that I even intended. It was like, you know, being in a foreign country—you know the language but you don't know any of the idioms.

Fortunately, the lessons he learned from that experience helped him to adapt in future career shifts. Years later he was successful crossing into government sectors and the medical field, and now feels competent communicating with many different kinds of scientists.

For the AI Specialist language hindered communication with relevant intellectual communities in terms of reading, writing, and interpersonal communication. His use of information resources extended beyond the sciences into philosophy, where he struggled with more than just the terminology particular to the fields. He felt that work in philosophy would have had a more profound influence on him if he did not have so much trouble reading philosophy texts, stating that it took him at least a year to read one book because the writing style was so foreign to him. He found writing for philosophers equally challenging.

> These people who I consider philosophers wrote this book, and they asked me to write a chapter on epistemic aspects of databases. I went to the dictionary and looked up epistemic. I could not figure out for myself what the epistemic aspects of databases were. After thinking about it for awhile, I figured out what I thought they must have meant and wrote a chapter and sent it off . . . but you see they had to suggest a title. I would never ever have titled anything I wrote something like that.

He had the knowledge needed to complete the task, but a shared vocabulary for talking about the problem was lacking. The title complication he described is at the heart of cross-disciplinary information retrieval problems. The knowledge documented in the literature cannot be tapped unless the information system and the user are using the same terminology or there are vocabulary switching mechanisms in place.

Face-to-face communication poses similar problems in the AI Specialist's research. Designing computer applications requires extensive communication with the groups of scientists that will be using the systems.

> When we ask them what operations do you perform on your data, their answer will be something like: we use a finite element multigrid method to do computational fluid dynamics. And they think they are answering our question. But, their answer is at much too high a level. Of course, we don't really know what that means. Although we found out our answer is at a much lower level: what bits of data do you keep where, where do you wish you had them, how many elements in your arrays?

On another project the information provided by the other group of scientists was too simplistic:

> When they think about programming, they think about it at a much lower level, like what kind of Fortran statements they are using and little details about their disk drives. So it's either this high or this low, and we need something in the middle.

There are obvious advantages to interactive communication where question and answer negotiation moves both sides into a meaningful conversation, but orchestrating a face-to-face situation where an exchange can successfully take place is no easy task. In this case, as with Tom's ambitious collaboration with the theoretical physicist, negotiating the level of discourse was assigned to a graduate student intermediary.

The Psycholinguist used linguistic creolization as an analogy for the language comprehension problems involved in interdisciplinary work, and this concept has also emerged in scholarly interpretations of how communication and culture evolve through discourse and interaction. Galison (1996) frames the creolization process within the construct of the trading zone, an arena where different disciplinary language groups coordinate diverse activities and subject matter. In his account of the emergence of the Monte Carlo technique in early nuclear bomb research, a provisional pidgin language matured into a "full-fledged creole," a hybrid language developed by a cross-disciplinary subculture that was able to support research without needing translation into a "mother tongue." "Pidgin" and "creole zones" are built as different professional groups develop intergroups and interlanguages (Löwy 1992). The Psycholinguist described the process in lay linguistic terms.

> It's not a real language; every speaker of the pidgin betrays their native language in the way they talk it. It doesn't really have rules and things like suffixes on words. It's very, very simplified. It's like the way Hollywood thinks cowboys and Indians talk. When I speak to a linguist or a computer scientist it's like this: me do experiment, get results, and they say, oh good—good experiment. We don't communicate the nuances; it's a simplified form of communication. I wouldn't tell a linguist or a computer scientist all the extra little things I would tell a psychologist and vice versa. We are communicating sort of rough ideas without all the rules.

He went on to explain that a pidgin evolves into a true language, or creole, within a single generation. The generation of children would be the students that are exposed to the pidgin; "they are creating a new discipline, the creole discipline, that people like me are never going to really appreciate." The creolization period spanned less than a generation in Galison's Monte Carlo case where the research community was small, localized, and focused on using computing to solve a specific problem. For the hybrids developing at the Center, disciplinary creolization may, in fact, be taking place, but the single generation notion is questionable especially since it involves synthesis between fields of inquiry that are global and distributed. And, there is no guarantee that the communication process will be greatly improved through the course of research interaction. For instance, Bowker (1993) gives the example of cybernetics, a generation of interdisciplinary effort that led to a "Tower of Babel"—a new specialized language that no one from any discipline understood.

STYLES OF THOUGHT

In his study of early 20th century German geneticists, Harwood (1993) identifies two styles of scientific thought. He categorizes scientists as either pragmatic or comprehensive based on their choice of research problems, their range of biological knowledge, and their interests outside science. The interdisciplinary generalists in this study have characteristics in common with Harwood's comprehensive scientific thinkers: they take a broad approach to problems, their attitudes support breadth of knowledge, and they strive for conceptual synthesis. The pragmatists considered the aim of science to be prediction and control, not unification. In contrast, and in accordance with the generalists at the Center, comprehensive geneticists "strived for breadth of perspective beyond their discipline" (257), and like a few researchers highlighted below, they had a fascination with form. Harwood refers to this quality as an aesthetic sensibility, and draws a connection between how artists and biologists think. Neil made a similar observation in relation to one of his interdisciplinary mentors who expressed his creative talents in various art forms as well as through his science.

Some of the cross-cultural communication problems described by the researchers appear to be related to styles of thought, or cognitive approaches to the scientific world, rather than terminology and context. When the Movement Specialist reflected on the time when he shifted into the field of psychology, he explained that he not only talked differently but also thought differently than psychologists. Recall also that Tom and the physicist needed their intermediary to do more than interpret words; the student was faced with reconciling two different ways of comprehending the problem that was being addressed. Tom acknowledged that the project moved slowly because "a physicist fundamentally thinks about the world differently than a chemist thinks about it." Neil made a similar observation when he attributed the Deep Space Nine atmosphere to the "physical types" trying to interact with the more qualitative biologists. But while interaction between researchers with different ways of speaking and thinking is problematic for communication, it can also be the key to interdisciplinary progress. Collaborations between differing mind sets may produce innovative results. Moreover, researchers who change and adapt to new disciplinary cultures during their careers develop unique blends of perspective, such as the Movement Specialist who brought insights from engineering and physiology to psychophysical problems. Becher (1990) encountered a good deal of what he called "occupational nomadism" in his study of research faculty, noting both risks and benefits of being nomadic that mirror the experience of interdisciplinarians at the Center. Changing fields is a risky step that can forestall career advancement. At the same time, gaining exposure to new knowledge and perspectives works to counteract the intellectual insulation that results from specialization. Alternative ways of thinking about problems can lead to innovative solutions.

Diverse career paths were the norm for researchers at the Center. Thus, it was surprising how frequently they described their work within the context of the discipline they had been trained in, rather than the one in which they were practicing. For example, the Computational Designer who studied physics in graduate school called himself a physicist, although he was working on computing problems and held faculty position in the computer science department. The Computational Neuroscientist, who was also trained in physics, said he thinks like a physicist, but works on biological phenomena. He described his research style as "a way of thinking about the world."

> I think in terms of equations and sort of quantitative descriptions of biological systems as opposed to descriptive biology, where you look at the thing, observe it, but you don't put it into a form.

Educational experiences have had a lasting influence on the researchers. Their research training has been internalized into their point of view, which raises important questions about the strength of personal association with accepted academic categories of knowledge. What background would be necessary for a scientist to see things from an integrated biophysical point of view? At what point will scientists be likely to state that they have a complex systems perspective of the world?

Form and pattern were a recurring theme in the researchers' descriptions of their cognitive approach to problems. A number of the respondents suggested that there are certain ways of thinking that are especially conducive to interdisciplinary work. The Device Physicist is multilingual and attributed his scientific thought processes to his years spent learning foreign languages.

> I tend to view things in the same way. I mean I may study different scientific topics, trying to see certain links, and then I study a different language and I see the same links. I don't tend to think about it. It is because the patterns, topology are the same. So actually, knowing languages helps me a lot doing science, because I use the same exact mental processes.

He also talked about his thinking in an active, visual way. When thinking about physics he creates "mental lines going through space"; he would be able to tell you "where the physics is going" by connecting it to a process but not the actual formula. More generally, the researchers conveyed the idea that interdisciplinarians must have a general "sense" of how things work or of what is "broadly applicable." The Biophysicist described it as the process of finding patterns.

> It is analogous to what you're doing in a computer when you do a search. You've acquired some knowledge or some data and it's here [points to her head]. When you are thinking about some new problem you are doing some sort of pattern recognition, a sort of matching thing, and suddenly you think, ah ha.

Along a similar vein, the Neurophysiologist, who latched onto the idea behind the cooperative work of insects, emphasized the importance of being able to find similar shapes or forms of activity that bear on problems that cross disciplines.

Pattern thinking, according to Anbar (1986), is a defining characteristic of a "bridge scientist," or a "poly-disciplinary" scientist who diversifies knowledge by transferring concepts across disciplines. Anbar identifies four basic types of bridge scientists, the most active and creative of which are generalists who "have become adventurers" and receive "satisfaction by stimuli from others who have different backgrounds and attitudes." This is a fitting description of the generalist researchers in this study. Bridge scientists of the second type are forced to cross disciplinary boundaries because an area they work in is no longer marketable. As we saw earlier, the computer scientists indicated that disciplinary obsolescence is part of their motivation for becoming more collaborative, and Sam claimed that most disciplinary problems had been solved. I found no commonalties between researchers at the Center and Anbar's other two types of bridge scientists—those with superficial training who attempt to get work or recognition as generalists, and those in appointed administrative bridge positions.

GROWTH CURVES AND CORE MAINTENANCE

> If you want to make progress in science, you have to be at the beginning of a growth curve. (The Photosynthesis Specialist)

Research trends influence the choices scientists make about which problems to investigate. There was an understanding among the researchers that if you wanted to "make it," you had to be active at the beginning of a growth curve, and that the disciplinary edges where hybrids develop are where new growth curves often start. For example, the Computational Designer talked about success as a "game" that for him is won by "playing at the edge" of engineering or biochemistry. Also, recall that Sam believed that the growth areas and opportunities were to be found in those dark crevices between the disciplines that had been obscured by the academic structures within universities. Jumping on a growth curve at the right time can be instrumental in obtaining research funding and advancing one's career, but researchers cannot spend all of their time and effort forging ahead in new territories.

> The other side of the science picture is that you have responsibility for seeing things through. . . . There is a lot of nitty gritty stuff that needs to be tidied up to complete a piece of work. So you get some of these little leaps up the growth curve and some plodding at the top of a growth curve. (The Photosynthesis Specialist)

Many researchers find that in order to play the science game strategically they need to sustain a firm position in a disciplinary specialization while they selectively target a particular interdisciplinary opportunity. As they explore new problems, they do not abandon their disciplinary concentrations, but instead maintain dual or multiple initiatives, building on a core research strength as they make the transition into a newer hybrid area. Core maintenance can keep a career intact and sustain funding while researchers start to work their way into a new territory. Tom learned about growth curve grant writing through his experience as an officer for a major funding agency.

> If it's too new you can't get money. When you are in the middle of things it's new enough that people are getting excited about it yet old enough to be mature enough that you can write a coherent proposal. . . . You need to have enough focus in an area to write a grant proposal that will be in the top 10 or 11 percent.

As the Computational Designer pointed out, when you explore new ground there is a risk of "disappearing below the horizon." Recognition and funding are key components of academic success, and they are easier to achieve in an established specialization than in a new area of exploration. Being an interdisciplinary scientist does not preclude the need to publish and gain approval in identifiable and mature scientific communities.

> It is sort of necessary to focus on one specific society and just create a corpus of publications there. Otherwise, people don't know you. You know, when you go up for a promotion you need to have a good number of letters from people outside and get good comments on your research. If you start publishing in scattered areas and in scattered journals you tend not to be very well known in your field. (The Device Physicist)

Keeping a career on track requires accomplishments along the way. In interdisciplinary work, the tensions between the purpose and standards of governance for research can interfere with career progress as well as scientific advancement. Alpert (1985) asserts that the "dissonance" between the goals and administration of science prevents interdisciplinary leaders from being recognized as "legitimate scholars by disciplinary communities or promotion committees" (261). There is a serious disaccord between what leads to a successful scientific project and what leads to advancement at a university. Tenure, Alpert contends, is contingent on individual achievement, which is judged solely on publication in refereed journals. Scientific progress, on the other hand, requires successful projects, which in turn depend on problem solving, collaboration, and innovation.

Publishing is the key to academic success, and there are fewer barriers to publishing specialized work within a disciplinary core. Work in new, emerging fields is hard to place and does not always fare well in the peer review process. We saw an example of this situation previously when the Neurophysiologist could not get his conceptual models accepted in the primary physiology journal in his field. The Human Factors Psychologist told an interesting story about two papers that he published one year in the same journal. He was reluctant to submit the first because it was a small study, that in his view, was of limited significance:

> We did this experiment as part of a larger project. . . . I assembled all of the data and there were a couple of coauthors on this paper, and I was looking through it and said to my coauthors, do you really think that we should publish this?

In contrast, he believed the second paper was pivotal—a new, solid scientific contribution.

> I'm very proud of the other one. I think it's a real accomplishment. I think it pushed forward the state of the art on a number of dimensions, not just one, but a number of dimensions.

Ironically, the first piece won the best paper competition for that particular journal that year.

> It is so interesting. It was supposedly the best paper in for that year, and I know it wasn't the best paper because I at least had one better paper, not to mention all of the other papers out there.

In this situation, the dissonance to which Alpert refers is exhibited in how the standards of excellence differ between the interdisciplinarian and the peer community that determines what is published and revered. The narrow, specialized work was appreciated and rewarded by the expert reviewers while the more innovative paper went unacknowledged. With this award, the specialized study that the researcher thought might not be worth publishing became more momentous than the progressive paper that was a step out into the research front. This has tangible outcomes, even at the local level, where external peer review plays a central role in the evaluation process. Awards are important assets for academic departments, and the implicit message to researchers is to produce the kind of work that will garner symbolic capital of this kind.

Interdisciplinary researchers experience the same "priority of paradigm" (Kuhn 1962) as disciplinary researchers when their new ideas are rejected because they are "at odds with the world around them" (Tom Tate). If, as Myers (1993) suggests, "lower level claims must be accepted by the rest of the research community of the subspecialty" before high-level claims can be made (340), then getting a high-level interdisciplinary claim published anywhere will be a much more substantial step than researchers anticipate. And, if a claim that is first introduced is too expansive, it has little chance of advancing through the benchmarks of scientific recognition. Claim making is an art that takes on a new dimension in the interdisciplinary realm where few are qualified to judge innovative approaches or novel findings.

The key to success for any scientist, according to the Organizer, is focus. In retrospect, he attributed the failure of some of the Center's programs to the tensions between individual success and more substantial scientific progress. Individual success requires focus, and, he warns, "you can only go so far in overcoming all of the sociological and intellectual factors that make for focus." People cannot afford to "reengineer their careers in order to explore an interesting new problem area." They have to find a way to make a new area fit. The Organizer noted that even the most interdisciplinary scientists have to make "cold-blooded" decisions about how to concentrate their energy, if they are going to be "hard-driving, ambitious scientists." Considering that most of the researchers need to maintain their core research track and at least one emerging area, their decisions about what projects to pursue and how to spend their time are considerably more critical than those working in a single core specialization.

A few of the researchers were specialists in areas that had helped them ease into a new hybrid domain. That is, they had been able to apply their core expertise directly to a problem that was ascending the growth curve. For example, the Perception Specialist developed techniques in the mainstream of visual attention psychology and then brought them to the interdisciplinary study of aging. He was then situated to do work that would be mutually beneficial to both areas of inquiry. Other connections take concentrated bridge work, as was the case with the Protein Specialist who linked his studies of protein structures to artificial intelligence, eventually advancing knowledge in two very diverse interdisciplinary communities.

An appointment at the Center does not alleviate the strains of a dual focus to the same degree as achieving tenure does. Tenure has the greatest impact on researchers' ability to change or expand their area of research. The Computational Designer stated it very simply: if you really want to answer general questions, "it takes much longer. The only reason I can afford to do that is that I already have all the promotions behind me." The Language Modeler agreed.

> Doing this at all is a real luxury that only tenured faculty can afford—to just, you know, take off and explore a new area, find out about new stuff, just hang out and talk to people, risk embarrassing

> yourself by asking naive questions. And the worst thing of course is to publish outside your field. Forget that if you don't have tenure.

In fact, he felt an obligation to help younger colleagues at the Center understand the risks of getting involved in interdisciplinary and collaborative work. But according to the Organizer the record shows that young faculty who are members of the Center are just as likely to get tenure as those who stay within their departments. One could argue, however, that since those accepted into the Center were in fact being acknowledged for important achievements, their chances of attaining tenure would be greater than the junior faculty at large.

CROSS-DISCIPLINARY TRAINING

> You shouldn't develop engineers who can do only one thing. (Sam Smith)

There is an understanding that scientists and engineers need to know more than in the past if they are to work in the hybrid areas intersecting the disciplines. It follows that interdisciplinary researchers should have broad scientific training, but most of the researchers at the Center were educated in traditional science programs. They amassed their broad and multiple perspectives by changing fields at various points in their education and professional careers, and in a few cases, through exposure to interdisciplinary role models. Boundary crossing work comes more naturally for the researchers with the most multidisciplinary backgrounds, like the Animal Learning Specialist who came to neuroscience after working in the hybrid area of physiological psychology or the Computational Neuroscientist who received his graduate training in physics but also apprenticed in an interdisciplinary science research program. Both of these researchers felt part of an interdisciplinary tradition, but the generation of researchers inhabiting the Center generally had little or no formal training in how to be interdisciplinary scientists.

The researchers are, however, acting as mentors for the new generation of scientists at the Center. Although it does not have an explicit educational mission, the Center is known to be an excellent training ground for young scientists and engineers. The Humanist praised the success of student apprenticeship at the Center, suggesting that the students constituted the primary force driving the path of the research. According to the

Organizer, "the graduate students benefit differentially more from the Center than the faculty do." They are exposed to opportunities they would never have in a traditional chemistry or electrical engineering department; they attend talks, work in multidisciplinary research groups, and have desks next to students from various departments. These experiences help them to "develop different attitudes" about how science is done than the students who are trained elsewhere in the university. The Computational Neuroscientist expressed a nondisciplinary disposition toward the students he hires. When someone comes to work for him from another lab at the Center, he "really doesn't know or care what the student's official department is"; it is more important that they have had experience in an interdisciplinary lab environment.

Some of the researchers seem to envy the students' freedom to work between the departmental "black holes," but others were sensitive to the difficulties students have negotiating their research-related social worlds. Like the professional scientists, students encounter serious obstacles in pursuit of interdisciplinary interests. One scientist explained that his students are much too busy taking care of departmental requirements to get much out of what other programs have to offer. The AI Specialist believed that it is more work, and a greater risk, for students to take on interdisciplinary problems. One of his students recently changed the direction of her dissertation work because she was uncomfortable in the amorphous space between disciplines.

> It was one of those cases where it falls right in the middle between two fields and nobody pays attention to it. She went to a conference and came back saying there was nobody there to talk to, so the second half of her thesis is more mainstream. She had to change it to get recognition.

In a later discussion with the AI Specialist, I learned that this student had been studying Internet security, and now that some time had passed everyone wanted to hire her. In this case the student did not spend much time on that precarious disciplinary edge; she was part of a quickly ascending growth curve at an opportune period.

It is possible for the Center to hinder a graduate student's educational experience and future success. Neil had seen situations where the lure of the Center compelled students to become too narrowly focused in their academic program. A couple of his advisees were drawn to the Center hoping to expand their scientific repertoire, but ultimately the computational opportunities that attracted them ended up limiting their base of learning. The demands of mastering multiple fields were daunting, and after a short time they dropped their biological studies and confined their thesis work to computational problems. Moreover, Neil found that once students went to the Center to work that it was almost impossible to get them out. Their positions gave them the opportunity to learn computational techniques and get involved in a number of projects, but they then became "consumed by the computational problems themselves" and relinquished the experimental part of their research program. Neil believed that these students would end up with fewer professional opportunities when they entered the job market. After graduation they would be competing with many exceptional computationalists, and, more importantly, they would always be dependent on experimentalists who were willing to work with them. Neil advocates training scientists who are skilled at experimentation and computation.

> You can ask your own questions and then you can answer them, and that is a big advantage. . . . In the future I would have to be stricter about saying no you can't go [to the Center] yet. You jolly well stay here and get some experimental results before you go up there.

As illustrated in the last chapter, Neil fully acknowledges the importance of computation for merging experimental and theoretical knowledge, but he has not figured out how students can develop both sets of competencies within the course of their studies. "The lure of the computer is very hard to advise against," he noted.

KNOWLEDGE LEVELS

For most of the researchers, knowledge development is time-consuming and difficult. Some of the scientists were very cognizant of their limited capacity to learn new material, especially at the level necessary to enhance problem solving. They felt the "burden of comprehension" inherent in cross-disciplinary work (Klein 1990), and the more subject areas a researcher spans, the greater the burden. Previous studies of interdisciplinarity indicate that integration across disciplines requires more than borrowing specifics from another field; researchers need to understand the surrounding context, history, and status (Klein 1990). White (1987) maintains that any meaningful crossing of disciplines "must take place by a process of translation that is based upon rather full knowledge of the practices that define each community" (11). Researchers must understand theory, technique, and particulars of the research problem, and therefore must know how to learn or find ways to merge new knowledge into their research process.

As Klein (1990) notes, interdisciplinarians need to know "what information to ask for and how to acquire a working knowledge of the language, concepts, information, and analytical skills pertinent to a given problem, process, or phenomenon" (183). Basically, they need to develop competencies that will allow them to deal with their increased burden of comprehension. But, as seen in the previous chapter, interdisciplinary researchers continue to rely on standard forms of communication. E-mail has changed the way they do business with their personal contacts, but the published literature, conferences, and personal networks are still the primary sources of information. The conferences and the colleague networks reflect how intellectual domains coalesce around research questions, but the strategies of accumulation are what make it possible for researchers to build a resource base for developing new knowledge.

How much do scientists need to know about another discipline in order to successfully incorporate it into their research process? Is it full knowledge of practices as suggested by White, or is it the analytical comprehension based on language and concepts described by Klein? Petrie (1986) proposes that researchers must acquire an interpretive level of tacit knowledge to do interdisciplinary work. They need to learn enough about the other discipline's cognitive map to be able to interpret a problem in that field's terms. In addition, "one can and probably must make this interpretive knowledge focal" (123) in order to learn it well enough to permit tacit functioning. Petrie based his notion of interpretive knowledge on Broudy's (1970) study of learning and Polanyi's (1958) concepts of tacit and focal knowledge. Polanyi contends that the "figure" in perception is understood focally and the "ground" is known tacitly. Petrie construes that "interpretive knowledge is almost surely used tacitly by the disciplinarian, and this explains why it is so easy to overlook its importance in interdisciplinary work" (123).

According to Petrie, there are two requirements for the minimum amount of knowledge to perform interdisciplinary work: understanding of the observational categories and the meaning of key terms used in the other discipline. It is questionable whether these criteria would in fact foster an interpretive level of tacit knowledge. In addition, these classes of knowledge do not match the various ways in which knowledge is developed and applied by researchers at the Center. Across the group, the researchers demonstrated considerable variation in knowledge of subjects outside their core, with their approaches to developing or gaining that knowledge differing according to research mode. For those with a generalist orientation, comprehension involves a high level of interpretive knowledge beyond understanding terms and categories. On the other end of the spectrum, interdisciplinary work is performed by collaborative teams in which individual members make contributions with less than Petrie's minimum level of knowledge in outside disciplines.

Detailed knowledge may be needed when researchers find that they lack a base of experience in other domains or an understanding of the history, status or context of the problem being addressed. For example, these excerpts from the AI Specialist refer to the specific practice-based knowledge needed for applied computing projects.

> [Computer scientists] in general don't know anything about information retrieval.

> We don't know anything about numerical analysis, so it's quite a challenge and it's really hard to understand what they do with their data.

> My students and I hope to do something to help them manage [their computing] and we know nothing whatsoever about what they do. . . . It took a really long time to get even a rudimentary understanding of it.

Each of these statements is tied to a different project that required significant understanding of the other community's orientation and work activities before the particular data management problem could be addressed. In one situation, the AI Specialist was able to get funding from a client group to send a student emissary into the field to do the necessary learning in situ. More often, the learning infringes on the researcher's own resources, time, and energy. Utilizing the core maintenance strategy, it is not unusual for researchers to "steal money from other projects" in order to absorb the cost of developing a knowledge base in another field.

High-level knowledge development is costly in that it decreases or slows down research production, at least in quantifiable terms. The Biophysicist explained that this was why she published fewer papers than other members of her department did.

> There is a cost to this, because there is an overhead associated with it. I mean you really have to know the context very completely of

> everything that you put out. You have to know what all the other literature is. You have to know how your work fits into everything else that has been done. So if you have to do this in three, four, five different communities, that's a lot more overhead for each paper that you put out.

The overhead expense is great for the Biophysicist and others with generalist tendencies who actively strive for a deep level of understanding in outside subject areas. Those working in a collaborative mode do not always work toward in-depth understanding, and therefore are not as burdened by comprehension problems. For example, the Vision Specialist had a very different perspective than the Biophysicist on the level of expertise needed in a peripheral area. He did not expect to be able to stay current on the literature of psychophysics or understand the complete context of research in that field. He maintained a "coarse level understanding of what characteristics matter" in adjacent areas of study. He talked about a kind of inferred imprint from outside fields. His knowledge in those areas was far from an interpretive level, but the other fields still had an impact on his approach to inquiry.

> These areas end up being more as influences that are under the surface when pursuing certain directions. So they are much more invisible guides, invisible motivations and inspirations than concrete acknowledgeable sources. I can't stand up and say I am doing this because of these specific observations, because that requires a lot of depth and study and also implies some sort of rigor in justifying why I am doing what I am doing. And that is not so easy because the knowledge about all this is extremely complicated, and we only know little.

This is a more passive posture to outside knowledge. It seeps into the research process, often through collaborative activities, but is not actively pursued. Where the Biophysicist would strive to develop an independent base of new knowledge, the Vision Specialist is dependent on others for their deep knowledge in areas outside his expertise. Thus he spends less time and effort on new learning and continues to be a productive scientist. In fact, he had been prolific, publishing more than 60 papers in a recent five-year period. Resigned to the fact that his knowledge of outside areas will always be limited, he takes what he reads and hears "on faith." The value of the information does not lie in his personal understanding. It is adequate for it to be accessible through someone else. This dependent, noninterpretive approach to knowledge may inhibit the degree of interdisciplinary permeation that can result. Where some of the scientists found other disciplines had a transformative effect on their research, the Vision Specialist expected other disciplines to have a subtle influence on his work.

The Computational Neuroscientist had a similar way of mobilizing knowledge from outside fields. In applying work from cognitive science to his research problem, he did not develop an interpretive level of understanding.

> If I was interested in finding out about models of attention within cognitive science, I could evaluate them within the context that I am interested in—whether that has anything to do with the way I think about my system. I can't evaluate them in the cognitive science context. You know, what do cognitive scientists currently think, or the strengths and the weaknesses. For that I would go down the hall.

In discussing his fabrication work, the Device Physicist appeared similarly removed from his cognate areas. He had managed to benefit from work done in chemistry without interacting with chemists or their literature.

> It's not that I don't understand the words. . . . I am not so concerned about understanding exactly what they are doing as long as there is a link in between which, you know, is going to feed the information. I just need the results.

He connected with chemists indirectly by finding other fabricators who had already built on relevant research in chemistry.

> I just need to know exactly what is important, and they will tell me what is important. It is a good way to collaborate in the sense that I don't need to micro understand everything. I just need to get the right amount.

His strategy was to locate where the knowledge has already been interpreted and applied in a way that was useful for his purposes. It was important for him to understand how the connection was made, but like the Vision Specialist, he more or less "takes for granted that they know what they are talking about" on the other end.

Neil had different criteria for what it means to know enough in an outside field of study. His necessary level of knowledge was directly related to the ability to communicate successfully with researchers in another community. Results alone are not sufficient—it is important to understand the language as well as the "meaning" behind theoretical perspectives.

> The difficult thing is you have to take the time out to become somewhat conversant with these new vocabulary words—the vocabulary of computational biologists and the vocabulary of theoretical physicists. That doesn't mean that you have to really become you know—I mean, I couldn't do a quantum mechanical computation, even the kind that they did on the backs of envelopes 40 years ago. This is just something which I never tried to do. But what I have done is try to understand the formulas and the meaning, if you like—the theories. And I think you have to do that in order to be able to communicate with the people who very often are injecting the novel ideas into the field.

This level of knowledge moves beyond what might be considered reading level knowledge to being able to have a discussion with a member of another field. For Neil, this is how real interdisciplinary integration happens—through asking, explaining, and discussing among multiple scientific domains.

As we observed earlier, communicating with an interdisciplinary audience demands that you understand how much a group has in common with you and each other. It is very difficult to make generalizations about the shared knowledge of any multidisciplinary audience. The Neurophysiologist, who does conceptual modeling, recalled a situation when he spoke to what appeared to be a homogenous gathering that, in fact, had a very small base of common knowledge.

> The first informal talk I gave to the complex systems group was a disaster. I started out talking about linear control systems theory and stuff like that. And like I say, I'm not a mathematician. So I thought these guys are all mathematicians and they are going to know what this stuff is, but they didn't. It's not that it was beyond them. It certainly wasn't, but it was a use of complex numbers that they weren't familiar with. . . . I didn't know that one group of mathematicians understood what another group didn't.

He ended up spending a one-hour presentation on the basic principles behind his talk. This frustrating experience taught him that with interdisciplinary groups it is necessary to "back way back to where any scientist can understand what you're talking about."

In contrast to the above cases, the Photosynthesis Specialist did not talk about the necessary level of knowledge in terms of influences, specific information, or communication. For him, knowing enough was a somewhat ambiguous stage of development. He defined it as being able to recognize when something is inconsistent with all the possible "overlapping hypotheses" in a problem area. In other words, a sophisticated "recognition of ignorance" leads to new discoveries.

> The most interesting things are the ones you don't understand. The trouble is that what makes things interesting isn't that you don't understand them because you are stupid or because you don't know enough—it's the things that you don't understand because you do know enough that they are not consistent with what you know. Those are the ones standing on their heads. If you recognize it it's usually because you know a fair amount about the system. . . . The trick is knowing enough to see when something is interesting for the right reasons.

For interdisciplinary research this idea implies the need for highly complex knowledge about a family of problem areas that may cross a number of fields of study. This type of knowledge is not dedicated to problem solving; it is an ability to look across fields to see what matters most and what is best positioned for investigation. Discerning opportunities for discovery hinges on a system of knowledge with the right levels and combinations of knowledge units. This perspective on knowledge development prevailed among the team leaders, such as Sam who felt that one of his primary functions as an interdisciplinary scientist was to identify the important research opportunities that lie among the established domains.

The series of cases highlighted here show how different work approaches are associated with varying levels of knowledge development in outside subject areas. The Vision Specialist and the Device Physicist, as collaborators, were not concerned with developing explicit levels of competence in other fields and had not been actively working to necessarily establish new connections between fields. They acknowledged influences from other disciplines but relied on being able to identify and apply connections that had already been made through the literature or through the assistance of colleagues or collaborators. As a generalist, Neil described a need for a deeper, more independent level of knowledge that can sustain meaningful dialogue between intellectual communities. The Photosynthesis Specialist described a more abstract type of knowledge that required enough depth and breadth to apprehend the complex of overlapping principles surrounding a problem and its potential implications. While he talked about his own goals and achievements, his role as a leader of an interactive team had no doubt been key to developing this unique type of knowledge base. When he sees something that is interesting for the right reasons, it is the result of a collective process of knowledge accumulation.

Nevertheless, it must be remembered that each individual interdisciplinary researcher maintains multiple knowledge bases. Some of these areas need to be more extensively developed, and some are more difficult to maintain. The Animal Learning Specialist thought of this situation as a literacy issue. "I sort of have to become an expert, if not an expert at least fairly literate, with respect to molecular biology." This area of competency required constant work for him because it was a topic that he had recently begun to incorporate, and the field was "exploding" with new knowledge. Neuroanatomy, another key subject that overlapped with his research, was a relatively low maintenance literacy because it was a much more static science in which he had built up a solid knowledge base before turning to more interdisciplinary research areas.

The knowledge strategy characteristics introduced in Chapter 1 pinpoint three dominant approaches to gaining knowledge as aligned with the various research modes. *Consulting,* favored by collaborators, is the practice of seeking information and guidance from knowledgeable sources. Collaborators work primarily on cooperative projects and tend to consult colleagues and other experts to capitalize on their expertise as needed. The *recruiting* method is indicative of team leaders whose primary technique is to bring knowledge to a project by enlisting specialists from a given area. Team leaders assemble research groups, attracting members who fill knowledge gaps in a problem area. Building one's own personal knowledge base is achieved by *learning*, the approach widely used by generalists who are less participative and more engaged in programmatic learning and long-term engagement with other fields of study.

The most versatile researchers are multimodal. They work both independently and collaboratively, consulting frequently with people in other disciplines and making full use of the diverse expertise available to them at the Center. They also take advantage of formal learning opportunities, such as seminars and workshops, attending meetings in areas that are far afield from their core for intellectual stimulation and to enrich their perspective. Interdisciplinary research requires a balance between established core knowledge and the infusion of new knowledge, but as the cases above demonstrate, the means and measures for developing applicable new knowledge are far from constant. Within the interdisciplinary modes, the researchers' main tendencies are to consult, recruit, or learn in order to build a knowledge base, but the magnitude of knowledge brought to bear on a research problem by any one person is widely variable.

OVERLOAD

> What biologists have been doing for a long time, well that is not powerful enough, it is not general enough, it's not sophisticated enough, it's not sufficiently broad enough to apply to these kinds of problems. Am I making sense? (The Protein Specialist)

Dissatisfaction with the limits of traditional scientific approaches is one of the forces driving interdisciplinary work. Interdisciplinary researchers want to be able to ask and answer bigger and harder questions. They want to meld the strains of science to do more "powerful," "general," and "sophisticated" science. This requires a command of more domains of knowledge and the ability to apply new or multiple methods. The overload problems associated with disciplinary scientific research are intensified in interdisciplinary research due to increases in the scope of the information landscape and the associated burden of comprehension, and the need to participate in collaborative projects and learning activities. In fact, many of the information practices discussed in Chapter 2 are strategies for combating overload. Consultation with personal contacts, attending highly specialized meetings, scanning bodies of literature, reading research reviews, and team learning are all devices for filtering out nonessential information.

Overload, according to Weick (1970), is related to the phenomenon of twigging, the fractionation of interests and knowledge that keeps the "need for knowledge in balance with the capacity to absorb it" (67). Resulting from researchers' need to keep up with information, overload is "the perceived inability to maintain a one-to-one relationship between input and output within a realizable future, given an existing repertoire of practices and desires" (68). Recognizing that most scientific work consists of idiosyncratic procedures, Weick emphasizes practices instead of skills, distinguishing three types of information work practices: processing, doing, and composure. Researchers process information using authorities, computers, research assistants, and memory. Writing grants and reading journals are "doing" practices, and composure work is devoted to attaining "presence of mind" and confidence. None of Weick's categories, however, accounts for the repertoire of information practices applied by researchers at the Center.

Interdisciplinary accumulation involves layers of processing, doing, and composure work, complicating input and output and exacerbating overload.

The processing and doing involved in keeping up with information clearly increased overload for the researchers. One of the AI Specialist's comments was typical: "I'm constantly falling behind and trying to catch up. I wonder if there are any areas in which I am up to date." Keeping current takes continual effort, and according to the general survey, it is perceived as mandatory. This situation causes a conflict between practices and desires, the two delimiters in Weick's definition of overload. What researchers think is necessary does not match what they actually achieve. In the survey, almost all the faculty members (92%) agreed with the statement "to be productive in my field, it is essential to keep up with the literature." At the same time, 72% agreed that "it is virtually impossible to keep up with the literature in my field." Also, 80% of these interdisciplinarians agreed with the statement "I have little time to read articles that are not central to my research interests." The AI Specialist exclaimed:

> I can't keep up with anything, even my narrow areas. . . . What other new areas I would read in I just don't know. The thought of having any spare moments for reading—I have huge stacks of stuff that needs to be read. . . . To even keep up with main trends is just not possible because there are so many more people than there were ten years ago and so many conferences publishing that are really pretty good. Before it was possible, but now it is not.

Weick asserts that managing overload leads to redefining the information environment. Relying less on reading and more on the knowledge of others is one primary strategy for reducing overload. "The more one believes authors and 'live' informants to be interchangeable, the lower the probability that doing and reading will be viewed as mutually exclusive, and the lower the probability that overload will occur"(89). We have seen a good example of this in the Device Physicist, whose primary way of knowing in peripheral areas was tapping the knowledge of personal contacts. The information environment at the Center supports a pool of "live informants" readily available for consultation. Generalists who accumulate broadly and do not branch into smaller units of inquiry are most threatened by overload, and we have observed that their need for the continual infusion of information and new learning can result in fewer publications. Overload leads to low productivity, which in the current academic reward system can mean denial of promotion or tenure.

By examining the conditions and circumstances under which interdisciplinary work takes place, we can now begin to understand how much of the research process is a precarious balancing act for these scientists. They need to accumulate information and develop new knowledge, but they must do so in a way that does not interfere with productivity or produce debilitating overload. They need to maintain a core research area while they explore new territories and get involved in new areas of scientific growth. Exchanges with other disciplinary cultures, that require learning new languages, methods, and norms, must proceed while a solid presence in one's official faculty unit is upheld. With so many demands on individual interdisciplinary researchers, the climate and policies of research organizations are instrumental in maneuvering any interdisciplinary enterprise.

---

[1] The end of certain kinds of scientific inquiry has been predicted in the past. For example, in the eighteenth century, the "Newtonian world-system was commonly seen as having extended the frontier of knowledge to its limit" (Fuller 1988, 274).

[2] Attributing the term to Lemert (1990), Klein (1996) defines shadow structures as the "structures and strategies that challenge the prevailing metaphor of disciplinary depth" (4).

Chapter 4.

ACCOMMODATING INTERDISCIPLINARY RESEARCH

As made evident by the information work, organizational structures, cultural conditions, and knowledge requirements involved in boundary crossing research, the scientists have developed sophisticated methods and routines that help them in their interdisciplinary endeavors. All of the scientists have found ways to gather information and communicate with researchers in other fields and overcome the departmental and disciplinary barriers that make it difficult to interact across disciplinary boundaries. To do this, some researchers rely heavily on the Center, and others have devised structures and strategies that are less entwined with the Center's resources or remain largely independent of the institution. It has been argued that the execution of interdisciplinary inquiry should radically alter the procedures of research and that it is insufficient to merely accommodate interdisciplinarity within existing structures or with fabricated bridge-building institutions (Fuller 1993). That proposition has merit as an ideal, but it does not help us to substantiate what is really happening in the work worlds of academic researchers where interdisciplinarity is not well understood and is rarely well supported. In essence, the Center is in the business of accommodating interdisciplinarity and buffering it from the deterrents ingrained in the system of academic higher education and research. Through a sophisticated plan and a large financial investment, it has managed to create a congruous microenvironment where the work has a better chance of thriving. At this point in time, well-endowed units like the Center probably offer the best possibility for success for interdisciplinary researchers working in academe. It is true, however, that this model does not radically transform the research enterprise in any fundamental sense. Rather, the experience of these researchers indicates that the progress they make, while supported and promoted by the Center, comes largely through their determined efforts to push forward in the larger network of institutions and cultures that continue to moderate science. Interdisciplinarity has yet to profoundly alter the processes that regulate research on a grand scale.

Institutions like the Center are rare. It is more typical for interdisciplinary researchers to carry out their day-to-day work within discipline-based administrative and social structures. This fact, however, does not diminish what can be learned from the practices of accomplished interdisciplinarians or from the successes of a resource-rich institution. The work done by the scientists and the environment proffered by the Center offer viable blueprints for developing interdisciplinarity within the context of conventional science and academics. Making room for researchers to pursue boundary crossing inquiry has as much to do with intent and design as finances and facilities. That is, a research environment can be built in a way that accommodates the balances that interdisciplinarians must achieve without significant investment in new resources.

One genre of research conditions plays a major role in advancing interdisciplinary work by creating leeway in the research environment, and to a large extent these leeway conditions can be induced in conventional research settings. The concept of balance is central to understanding how the scientists at the Center exploit institutionally created leeway in combination with their own research strategies to bring order to opposing forces in the research environment. The biophysicists stood out as the most expert interdisciplinarians in their skillful negotiating of dichotomies. Their adaptations to local and global disciplinary dynamics and their processes for addressing research problems are exemplars of interdisciplinary practice. The stabilization function of the Center and the versatility of collaboration further accommodate researchers' boundary crossing activities. Together, the balances achieved by researchers and the solidity produced by research structures permit interdisciplinary approaches to operate in a disciplinary world.

CREATING LEEWAY

The Center as a whole was not a naturally emergent network of researchers. It was constructed out of groups of associated faculty that were selected based on their proposed or ongoing research problems. The problems they were addressing were deemed important, and it was predicted that work on those problems could benefit from dedicated infrastructure and additional resources. In her case study of cancer research, Joan Fujimura (1987) develops a model of how "do-able problems" are constructed in one area of medical research. Her framework is organized in three levels: the experiment at the bottom, the laboratory in the middle, and the overarching social world at the top. Answering a research problem is dependent, she argues, on the successful alignment of all three levels. This doable problem structure is useful for interpreting research work at the Center because, as we have seen, the researchers are problem oriented and the research groups designated by the Center are based on current scientific problem areas. In essence, the Center's role is to make interdisciplinary science more doable. Moreover, individual researchers continually strive to make their problems more doable by employing strategies that help them work across boundaries.

The Center influences the research process at all three of the levels identified by Fujimura. It provides facilities and material support for doing experiments, running labs, and participating in social world interaction, and the researchers have acknowledged the institution's capacity to make a difference in how research is done. There is a sense that the Center has a holistic effect on research—it is a place where science is "done right" or the "the way it is supposed to be," and the high level of material support is a significant part of this sentiment. The Humanist commented on the inequities between the Center and his home department on the other side of the campus. As he suggested, "who gets to play the game" is an interesting question, since the resources at the Center are among the most extensive at the university.

Fujimura (1987) asserts that abundant resources provide the "leeway" needed to construct and solve scientific problems, emphasizing the importance of the outlay devoted to the administration of work and the division of labor. The coordination of scientific labor is especially cogent for interdisciplinary problem solving because of the complications inherent in managing teams of diverse scientists. But the concept of leeway needs to be extended to incorporate allowances that are not financial or material in nature, but are nonetheless essential for managing scientific approaches that differ from the norm and depend on meaningful interactions between communities. Emanating from various organizational and individual initiatives that lie largely in the social worlds of science, interdisciplinary leeway is a nontangible resource that makes room for or yields the freedom to do boundary crossing research work. Researchers are limited by finite amounts of work time and individual knowledge capacities. To manage the extra demands of boundary crossing research and continue to be productive scientists, interdisciplinary researchers need mechanisms for producing leeway built into their work situation

The Center creates leeway at the experiment and laboratory levels by channeling resources to articulation work (Strauss 1988, Corbin and Strauss 1993), thereby shifting the burden of this type of work away from the researcher. According to Fujimura, articulation work—the planning, organizing, evaluating, adjusting, coordinating, and integrating activities conducted by research teams and their instrumentation—underlies successful science. Expertise and materials are needed to take care of the seemingly mundane processes of organizing and reorganizing the work of managing labs and experiments. Fujimura notes that leeway is increased as articulation is streamlined through the division of labor and by packaging and standardizing tasks. But, she also warns that standardization can work against novelty. This could be a serious tension for interdisciplinary initiatives, which generally strive for inventiveness. We have seen a number of instances of streamlined articulation that do not appear to be inhibiting innovation, such as the journal club's systematized literature searching and the more general use of collaboration as a method for dividing project work. With the layers of boundary work involved in interdisciplinary science, the management of articulation work becomes more complicated and crucial.

The Center provides high-level support for both explicit and implicit articulation work (Strauss 1988). Some examples of explicit, formal articulation services offered in-house include: a department for proposal writing, grant administration, and travel arrangements; a unit responsible for conference and program planning, materials, and publications; a high-tech, service-oriented library; and state-of-the-art computing and visualization facilities. Implicit articulation takes the form of secretarial support and abundant office and lab space. In the general survey we established that researchers experienced significant benefits from greater administrative support and quality laboratories. The full effect of this extensive explicit and implicit articulation support is not known, but in the survey researchers who had been affiliated with the Center for a number of years, and who spent a good proportion of their research time there, reported substantial increases in their productivity. Through the subsequent interviews, however, this finding became ambiguous since productivity emerged as an area of concern for both researchers and the administration, in spite of the improved base of support and perceived increases in research production.

In Fujimura's model of doable research, the social world is by far the most complex sphere. Based on the way that the researchers talked about their work, there is no doubt that the social world requires the most leeway support. The experiments and other lab related work were taken for granted as routine and relatively unproblematic compared to the activities situated more directly in the social world. Breaking into new disciplines and communicating across communities were the laborious and difficult problems that did not have financial and administrative solutions. Boundary crossing activities compound the number of complex social worlds a researcher must align during the problem solving process. As Fujimura proposes, much social world research work has to do with communication—the tasks involved in surveying literature, corresponding with and talking to other scientists, writing articles, and the other practices aimed at conveying how your experiments contribute to a research area. Convincing colleagues, enrolling "allies" (Latour 1987), and building research networks (Callon, Law, and Rip 1986) are social world processes that gradually bring together all the constituents involved in an interdisciplinary outcome. For example, importing and exporting information are acts of alignment, as are the translation activities performed by intermediaries and the team building efforts undertaken by laboratory managers. These are social world processes that calibrate resources and relevant knowledge from diverse disciplinary domains and channel them toward interdisciplinary problems. Fujimura's framework can be readily applied to the interdisciplinary process primarily because it gives adequate emphasis to the cross-cultural, social world dimensions of boundary crossing research.

Many communication-based alignment activities have been actively attended to by the Center. It has supported wide-scale exchange of ideas by organizing national and international workshops and conferences and by bringing in visiting researchers from other universities, industry, and government. Sophisticated electronic communication systems were put into place to promote interaction between distributed researchers. Locally, the facility has common spaces among offices and labs and ample formal and casual meeting areas. Ongoing seminar programs have made knowledge development more convenient, thus reducing individual overhead for learning activities. Literature overload problems have been addressed by library services that support cross-disciplinary information gathering through individualized current awareness and document delivery programs. The library distributes customized lists of citations to researchers based on their topics of interest, and all materials that researchers select are delivered to them.

Nevertheless, even with its multitude of resources, conscious design, and well-placed efforts, there are limits to what a building and administrative services can achieve at the social world level of science, especially when the most coveted and critical resource for the researchers is time. For instance, the Psycholinguist expressed genuine disheartenment at not being able to learn more about complex systems research, even though this area is actively supported by administrative services.

> I've gone to two or three talks, that's it. Enough to be intrigued. It would take a lot of learning. I mean, my ignorance of that is tremendous. It would be very, very hard to do. I would have to stop doing other things.

Time is the scarcest resource, and thus creating time, so to speak, is perhaps the most valuable contribution the Center can make for researchers, at least at the individual level. With its centralized administrative functions, the Center has been effective at reducing some time constraints through the successful translation of resources into articulation work. Just as importantly, the Center acts as a commons, a gathering place that increases the opportunities for interaction with complementary colleagues. Physical proximity "creates" time through convenience. Researchers commended the Center's role in making interaction "more convenient" and making it "happen in a shorter time scale." For the full-time researchers at the Center the proximity factor is paramount. For those who continue to work primarily out of their home departments, the Center is less essential but still useful as a place to talk shop and attend seminars. Most of the department-based affiliates found the Center to be too far away from their primary work site, and apparently, there had not been enough incentive to physically move their base of research there.

Proximity Leeway

The Center has not ended up to be the daily work location for all the researchers, but it has played an important role in the intellectual life of most of its members. Intellectual proximity has created more influential leeway than physical proximity. Physical nearness to other scientists can enhance the research process, but as one researcher stated, "it takes more than just physical presence. . . . It doesn't just happen because you put them all together in a building." In fact, many basic interdisciplinary research practices do not require physical proximity, especially now that networked information systems are commonplace. The researchers' collaborative relationships are perhaps the best example. The location of Tom's collaborators varies from project to project: some partners are seated at the Center, some are members of different departments on campus, and others work at universities in other parts of the country. After coming to the Center, researchers continued to pursue projects with scientists outside of the Center and the university, and these distributed groups usually constituted the scientists' most important reference group.

For the Computational Designer, the Center's physical presence was of little consequence, at least in terms of collaboration.

> I was already collaborating with people before the Center, and I don't actually think it has made any difference for anybody. I think that people were already collaborating or not. I don't think that by putting people in a building together they start collaborating.

The general survey showed that 75% of the researchers had conducted joint research with individuals in other disciplines prior to coming to the Center, while only 69% collaborated with individuals from other disciplines since joining the Center. The majority of the collaborations that the scientists talked about grew out of contacts made at conferences and through their personal network of colleagues apart from the Center. Taking into consideration the difficult work and the barriers involved in boundary crossing research, this lower percentage is not surprising, since about a third of the respondents had been affiliated with the Center for less than three years.

Local partnerships will likely increase over time, but the researchers' strong relationships with distant colleagues demonstrate that it is not necessary to have collaborators at the Center to perform interdisciplinary work. Having people nearby to talk to, on the other hand, is one of the most appreciated features of the Center. Accordingly, the most difficult aspect of the research process for some department-based scientists is their outlier status; they do not have the leeway created by convenient access to peers who understand and value their research. Center-based researchers have colleagues available for consultation or collaboration. They share a sense of place and have things in common with each other, although the commonalties may not lie in their disciplinary orientation or training but in their problem areas and problem solving approaches. They also share an understanding that they would not find this type of kinship in a more conventional academic setting. Several researchers had experienced isolation in research organizations in the past, and even those with interdisciplinary backgrounds and high stature in the world of science were grateful for their new niche.

> The Center sort of validates this approach because what I am doing doesn't overlap with any of the traditional disciplines extensively. I get pieces from each one. So if I were in a more traditional environment it would be harder to find like-minded people to sit down and talk with. (The Computational Neuroscientist)

Even Sam, an internationally acclaimed scientist, recognized the importance of being part of a locally recognized research community.

> When I was in an academic department I was always at the edge of the field. Now I am at the middle of the Center. Everything I do is just right for the Center—the middle, the center. It is nice to feel that way. It is not so nice to feel always on the edge. It is actually comfortable to know belonging here.

The edge is a precarious place to be situated, and being on the margins alone is much more difficult than it is as part of a group. The researchers do not need to be located near their collaborators, but they do need intellectual comrades in their daily work environment. According to the Biophysicist, the Center has put together a "stew of really disparate elements" to produce a functional pool of "creative and atypical people."

Symbolic Leeway

The researchers who did not find any advantages in congregating at the Center still understood the role it played symbolically. The validation felt by the Computational Neuroscientist in his statement above is an example of the symbolic leeway generated by belonging to a prestigious research organization. The Movement Specialist, who claimed few benefits from the Center, admitted that it gave him "a foot in the door to get people to listen . . . a certain measure of credibility in places where I really might not have had the credibility." He was working in a relatively new research area that had grown out of a field that was "held in low esteem in the scientific community" in the past. His association with the Center counteracted the residual influence of his low status disciplinary origins. Researchers who are part of established, high esteem fields also feel a need for symbolic leeway. Sam was grateful for the prominence of the Center, noting that when he first took his position a few years ago, interdisciplinary research had a bad name.

> Not long ago, in my field a biophysicist is a guy who talked biology to physicists and physics to biologists. In other words, they talk to the guys they don't know and try to appear good. Today I can laugh about this because I do not hear this joke any more. The Center is a fantastic symbol—a very well endowed research institute. People like to be around it. It's prestigious because it looks so nice. Center researchers are now connected with something very positive.

An appointment at the Center makes it legitimate to continue on a research track that has been dismissed by a traditional discipline, as it did for the Neurophysiologist whose work had been ignored by his core community. As a Center affiliate he had the authority to pursue his theoretical modeling work even though it continued to be discounted by many of his peers.

The researchers did not need to physically move into the facility or be active in their designated Center-based research groups to benefit from symbolic leeway. The fact that they received a position there meant that they had been recognized as a promising interdisciplinary researcher. The appointment provided an impressive entry on their curriculum vita, and it was influential in attracting collaborators, funding, and future employment opportunities. At the personal level, however, the researchers who physically gravitated to the Center profited from symbolic leeway, in the form of prestige, and proximity leeway, in the form of colleagues. Moreover, they tended to be more content with their situations than the affiliates who had the symbolic advantages without the intellectual support.

Technological Leeway

The Center is known for its strength in computational research, and it serves as the headquarters for computing activities for many researchers, including some who continued to work mainly in their home departments. The methods used by these scientists require powerful and sophisticated computational technology, as they push the edges of their fields by continually applying bigger, faster, and more complicated computational techniques. Consequently, research questions at the technological edge are only doable when there is a workforce that is skilled at using computing applications and that can continue to upgrade and develop new computer methods. Computationalists at the Center were not concerned with obtaining and maintaining equipment for their research. The infrastructure was in place, and they had access to state-of-the-art facilities and strong support staff. However, they did not necessarily have complete technological leeway. Computing was perceived as a "time pit" that consumed precious research hours, the rarest of resources. The lack of time was one main reason researchers gave for not taking advantage of electronic information sources such as databases or electronic conferences. When setting priorities for learning new technologies, computational methods took precedence over information technologies, even though the learning curve for information systems was likely lower. Some individual researchers, such as the modelers, maintained their own personal computational expertise; others achieved technological leeway by using a participatory approach. Since few of the researchers could afford to devote the necessary time and effort to developing the technical know-how needed for their entire program of research, computational colleagues had become extremely valuable commodities. The Structural Biologist revered his software designer colleague who acted as a "window into computer science."

Large lab operations can build the necessary computing expertise into their work force. Sam insisted on a high level of technological competency within his group. Computing, he explained, "can speed up what people can do by hand." He takes great care in selecting students to work on his projects, looking beyond their qualifications as mathematicians, physicists, and biologists, to find people who have grown up with computers. In his view, the students with real potential have had computers as their "hobby"; those who do not have the "cultural technical background" are a "losing battle." They may succeed in picking up some skills, but according to Sam, they will always trail behind. "The fact that we have these abilities in many of our students helps us create the discipline of computational biology." Sam is very interested in moving forward quickly, and therefore students for whom computing is second nature are a valuable leeway producing asset.

BALANCING DICHOTOMIES

As we have considered the interrelationships within the research process, we have observed how domains intersect within problem areas, collaborations, laboratories, and the literature, and how members of different research communities interact through boundary crossing work practices. The networks of interdisciplinary research, traced through the sites and activities of the scientists, function despite many constraints. Researchers develop stases that balance their orientations and efforts, counteract opposing forces, and stabilize the complex relationships of interdisciplinary research. These equalibriums are most pronounced in the work of the biophysical researchers'[1] a group that has succeeded in forming constructive balances in their research processes. The intersection between biology and physics, referred to by one scientist as "new physics," is a mature hybrid area that sustains an active flow of people and knowledge back and forth between communities.

The biophysicists are aware that an "intellectual void" still exists between the two domains. One biophysicist remarked: "Physicists aren't interested in anything larger than electrons or smaller than the universe. Between cosmology and elementary particles there is this sort of intellectual wasteland." The link between the two disciplines is established, yet there remains a clear difference in perspective and technical expertise. The outcomes of new physics are functioning in the scientific enterprise while physicists and biologists continue to assess and sift each other's methods for analyzing the scientific world and solving biophysical problems. Reciting a favorite quote from an anthropologist, Neil questioned the notion of physics as a "hard" science.

> Well, you know "there are the hard sciences like physics and then there are the really difficult ones like paleoanthropology." I think that that's really an important distinction which many people don't realize. Physics of course is hard. It is a hard science in the sense that it is well founded and has a very firm foundation. But in many ways, and I've seen this happen over and over again, it's just as hard and maybe harder for a physicist to come in and make sense of biology than it is for biologists to make sense and acquire skills in physics.

The backgrounds and career paths of the researchers at the Center were mutable. Several had moved from physics to biology, beginning their work on biological problems after they had schooling and training in physics, and vice versa. The physicists were drawn into biology because some of the most compelling scientific problems are biological in nature, and the biologists were attracted to physics because their research problems can be addressed with sophisticated physical methods. Expertise with quantification and formulation, which is important for theory development with computer modeling, is developed through the study of physics. The study of biology develops an understanding of the complexity of life. Progress in biophysics comes through a balanced application of approaches, methods, and knowledge from both orientations. The work of the biophysicists showed how different dimensions of science can be moderated and kept in balance in the conduct of interdisciplinary research.

Experimental/Theoretical

A number of scientists talked about the importance of both experimental and theoretical work to their area of study, and the biophysicists, in particular, placed as much value on work that integrates theory and experimentation as they did on interdisciplinary contributions. Only one researcher thought of himself primarily as an experimentalist, and others identified themselves as theoreticians after having devoted years to experimental research. In her field study of a nuclear physics laboratory, Traweek (1988) identified a divide between theorists and experimentalists within a culture where young physicists learn the "rituals of avoidance" before they learn how to "forge and maintain proper links" between the two orientations (113). This type of divide was not evident within the group at the Center, perhaps because there were so few classic experimentalists. Biological and physical scientists, modelers, and an engineer all suggested that there is no substitute for the insights gained through concentrated experimentation. The Neurophysiologist admitted that the intuition that allowed him to now do "pattern recognition" across disciplines grew out of his extensive experimental experience.

An interesting nuance to the experimental/theoretical relationship was introduced in the last chapter. For some students, experimental training was being displaced by the development of computational competencies, and Neil was very concerned about the prospects for students who abandoned the experimental part of their education. The attraction of computational programs of study was working against the experimental/theoretical balance. If this shift in emphasis were to become widespread, a shortage of experimentalists could impact science more generally by creating a situation where modeling work escalates without the infusion of new data. There will continue to be a need for experimental data, and therefore experimental expertise, in the scientific work force. The question remains whether individuals can achieve the necessary abilities to be competent scientists in both areas. The Neurophysiologist talked of working long into the night modeling data gleaned from journal articles, but he was understandably more excited about a prospective collaboration with an experimental physiologist who "would know the facts of the system as well as anyone in the world. . . . That's his life—those facts."

The biophysicists have developed a variety of research practices and information routines that are aimed at maintaining the experimental/theoretical duality. Some construct research teams that include both types of scientist, and others give due attention to theory by identifying and interacting with good theoretical communicators. A few researchers mentioned the importance of reading popular science books written for the more general public. General and derivative texts are also studied and consulted for theoretical basics, and personal contacts are relied on for talking about how theory relates to specific questions. The researchers frequently mentioned the importance of multidisciplinary journals like *Nature*, *Science*, and *Scientific American* that introduce new approaches and current theory in peripheral areas. On the experimental end, conferences are the best venue for keeping current with the "raw data" and the new results.

Collaborative/Generalist

Among the biophysicists there is a preponderance of scientists who follow the generalist mode but who take on collaborative projects on a fairly regular basis. There is a second, smaller group of psychologists who have a similar profile. These collaborative generalists have strong ties to the Center. They are full-time affiliates who have readily adapted to their new environment and are profiting from the centralized location. Like all the scientists studied, they bring multiple disciplines to bear on the problems they investigate, but their multiplicity is more pronounced than the other researchers because of their dual research modes. They employ generalist and collaborative approaches, engaging in some group research while placing considerable emphasis on individual learning and on expanding the scope of their program of inquiry.

Compared to the more exclusively collaborative or generalist researchers, these bimodal scientists use the greatest variety of information gathering practices. They are highly consultative, making use of the expertise available to them at the Center, and their knowledge base development is deliberate, directed, and part of a continual cycle of learning and probing. These are the researchers who take advantage of formal seminars and workshops at the Center and at conferences, and who teach to learn and self-teach through broad reading and intensive one-on-one interaction.

One biophysicist's generalist orientation was strongly represented in his beliefs and practices. He felt that to do good interdisciplinary work one must engage with relevant peripheral subject areas beyond the "superficial" level. He relied on reading and consultation for knowledge development in his cognate fields—a combination of practices common to many of the generalists. One of his primary collaborative strategies was recruiting research associates, primarily graduate students, who could contribute pure computer science expertise. In terms of productivity, it is possible that he could have achieved more technological leeway by working with trained computational biologists instead of computer science students. A strictly collaborative researcher might have preferred this approach. However, more applied computationalists would have contributed less breadth of perspective, the feature that made his work a significant interdisciplinary accomplishment.

It is not possible to determine if there is a causal relationship between being based at the Center and having this dual orientation. It seems most plausible that those who tended to be bimodal have thrived in the new environment, and shifts in the population of the Center as it stabilized could have produced a greater concentration of collaborative generalists over time. Beyond that, when the institution was being established, being "interdisciplinary" was a criterion for placement but there was never a guiding definition of what it meant to be interdisciplinary, leaving the selection open to individuals who were highly collaborative as well as more individually integrative researchers. Once a scientist was established at the Center, the leeway produced by the resources and convenience of interaction may have made it easier for individuals to meld these two orientations. Researchers who already tended to be collaborative had more opportunities to broaden their personal knowledge base, and those with a generalist orientation had an attractive new pool of potential collaborators.

There seems to be a continuum within the collaborative/generalist approach that is roughly aligned with production levels, at least in terms of publication output.[2] The researchers with the greatest number of publications, the Vision Specialist and the Structural Biologist, were among the most collaborative. The highly collaborative group also included the Device Physicist and the Computational Designer, who did not appear particularly prolific according to quantity, but who explicitly stated that they had managed to publish more than their departmental colleagues. Those who had the fewest publications, the Human Factors Psychologist and the Movement Specialist, clustered at the generalist end of the continuum. The collaborative generalists had more moderate publication records. Their ability to complete a respectable number of projects while continuing knowledge base development activities suggests that their dual orientation may be instrumental for balancing focus and breadth.

Focus/Breadth

The difficult equilibrium between focus and breadth revealed itself in the respondents' concerns with productivity. From the Organizer's perspective, productive scientists must be focused scientists, and the researchers were keenly aware of the conflict between breadth and production. Tom Tate expressed a common sentiment: "I've got so many things going on right now I can't keep focused very well and that's not helping productivity." Somewhat surprisingly, there was no indication from the researchers that the act of collaboration necessarily increased their productivity, and in some instances it did the opposite, increasing workload but not output. Of Tom's collaborations, the most demanding one had yet to produce any results. The need for extensive new learning and translation had kept the project from advancing, but despite the slow pace, Tom was quick to acknowledge the value of the process. A collaborative project that spans broad domains may be painstaking, but it can also offer a unique opportunity for knowledge base development.

> Where you have the least in common you learn the most because you are stretching yourself more. On the other hand, for productivity, you are far better off working with somebody you already can work with. . . . If there is one person way over there and another person here and they are trying to find common ground in the middle, well, sometimes it works and sometimes it doesn't. But in trying to get to that common ground, you are covering a lot more territory.

Gradual, strenuous collaborations are a luxury that bimodal researchers may be able to afford if they are also engaged in other more short-term projects and there is adequate leeway built into the research environment.

One biophysicist, whose research bridged the domains of protein structure and machine learning, had achieved a functional balance of breadth and focus and his boundary crossing efforts had been influential in both molecular biology and artificial intelligence. Although crossing over into computer science continued to be difficult in terms of information gathering and comprehension, he had managed to publish regularly and gain considerable recognition in both areas. The Center succeeded, in his view, at making interdisciplinary research happen more quickly and conveniently, but the intellectual work of integrating the two areas remained a significant struggle. Over time, he has been able to learn enough about computational tools developed in artificial intelligence to successfully design new ones specifically for the study of proteins. Finding time to learn was a major complication, and, not surprisingly, language was one of his greatest barriers. The machine learning literature caused him the most difficulty, since it has been written "by computer scientists for computer scientists" for decades. Filtering through the abstract words, ideas, and jargon was time consuming and frustrating work.

> Half the stuff was just defining new words or coining new words that would describe a particular feature. It was trivial. And among each group of researchers there is no consistent nomenclature or definition.

He is not alone in his complaints about the computational discourse. Other researchers, including a computer scientist, also expressed dissatisfaction with the inaccessibility of the computer science literature.

The focus and breadth equilibrium is related to the core and scatter relationship introduced as part of the conceptual framework in Chapter 1. Scatter creates the breadth that takes researchers away from their focus or core. Knowledge is concentrated in an intellectual core, and, at the same time, overlaps in the periphery through scatter (Chubin 1976). This is essentially what the co-word maps detailed for each researcher in the first phase of this study. Interdisciplinary research places more emphasis on the peripheral areas than discipline-based inquiry, resulting in a greater degree of balance between core and peripheral knowledge. The interdisciplinary research process accumulates peripheral knowledge around a core knowledge unit. Researchers' activities channel dispersed knowledge toward a specific problem, reconfiguring the core by reinforcing and initiating links to relevant peripheral areas. For many scientists at the Center, their core is already a mix of disciplines—a hybrid; and in some cases these hybrid cores are highly specialized research areas. For example, the Vision Specialist worked on very specific computer vision problems by drawing from less specialized peripheral domains of psychophysics, neurophysiology, and graphics.

The core maintenance activities described in the last chapter help keep researchers grounded in their focal knowledge area. Accumulation strategies make it possible for them to capitalize on the periphery to build a broader and more powerful base for understanding and investigating a problem. We have seen that knowledge strategies affect individual levels of core and peripheral knowledge and that specific types of leeway allow researchers to shift their efforts away from the core. Abundant resources, rewards, sense of community, validation, and technological capability accommodate ventures into the periphery. Unplanned events and unexpected discoveries can also steer scientists toward new paths and alter the focus of their research.

The researchers were forthcoming about the role of serendipity, happenstance, and coincidence in their work. The Human Factors Psychologist compared science to dating. "It's like meeting someone in a bar"—connections are often made by chance. It is also evident that these researchers employ strategies to increase the chances of serendipitous discovery. For example, those who practice undirected broad reading and attend talks that are far afield from their core are engineering situations where fortuitous discoveries in the periphery are more likely to occur. Scientists may also shift their investigation away from their initial focus during the course of experimentation. In one case, a biophysicist accidentally disproved a hypothesis about how plants adapt to the environment. During an experiment, he recognized that his data was inconsistent with his understanding of a certain biological system, and he had enough grounding, or balance, in a peripheral area to see that the data conflicted with his assumptions. This discovery put him on a completely new research track.

CENTRALIZING HYBRIDIZATION

The Center proved to be an advantageous alternative structure for a number of the distributed groups that had previously existed within the university. It was especially valuable in its role as a commons. The Device Physicist stated that general brainstorming did not really happen very frequently among local scientists, but when it did it was likely to take place at the Center with colleagues you knew and worked with every day. Of all the researchers, the collaborative generalists made the most use of the Center as a home base for their daily activities. It supported their bimodal approach, serving as a central location for cross-disciplinary interaction *and* knowledge base development activities. This group flourished in the research atmosphere created by the building and its programs. For some, like the Language Modeler, the most important aspect of the environment was the effect of the diverse stimuli collected under one roof. Together they resonated something intangible but penetrating.

> Things just kind of creep into your head without your thinking, oh this is something new I have to know about, or I'll file this here. It just kind of gets stored and pops up sometimes. I mean, I guess I'm just here being a sponge in a sense.

The Humanist was similarly struck by the Center's indirect influences that "buzz in inaudible frequencies." For him even the posters and fliers were important to the process of intellectual assimilation.

The stimuli afforded by easy access to people from different disciplinary backgrounds also promoted hybridization. The Center provided an excellent place to consult with scientists in cognate areas and to talk shop. As Brown and Duguid (1991) have shown, sharing "war-stories" is an important part of the "periphery of communication" where learners pick up valuable "information, manner, and technique" (50). Interdisciplinarians are experts, but they are learners as well, and they thrive on the ability to have casual conversations with colleagues they would not have contact with in an ordinary academic position.

> You just wouldn't run into them in a physiology department—thinking wild computational geometry things that might have applicability to an area like computational biology, or thinking of ways to put processors together in parallel to do much bigger simulations than you could otherwise think reasonably about doing. So the Center really provided a lot of people to interact with who just wouldn't be hanging around my department, and they just really make big contributions. (The Biophysicist)

The researchers reported that working at the Center increased opportunities for small group and one-on-one discussions, the type of interactions that are valued for increasing knowledge and triggering new ideas. However, impromptu dialogue happens less often that we might think. As an observer from the outside, I did not see a high level of informal conversation at the Center, in part because during my years working out of an office there I did not have access to the laboratories, only the public areas and the offices of respondents during interview sessions. At the outset of the study, I had expected to see a flurry of science and hear a hum of brainstorming and discussion. I was continually surprised by the stillness of the building.

It is the informality of the dialogue among the scientists that is significant, not the frequency. In the researchers' descriptions of their interactions with others, they rarely discussed their associations within the twenty official, administratively designated units of the Center. The important exchanges were casual and took place within their laboratories, between project team members, or with the person "down the hall." These relationships regularly cut across the formal research groupings. A number of the collaborative generalists were very conscious of their cross-group unions and the metastructures that were forming. The physical centralization of cognate researchers succeeded at making cross-disciplinary interaction convenient, which led to emergent connections that were more fruitful than the institutional delineations. This primacy of the informal concurs with Brown and Duguid's (1991) findings that administrative charts and other descriptions of organizational relationships are usually fundamentally different than the way people actually work.

A central facility appears to be key to the discipline-building function of research organizations identified by Lenoir (1993). The Center supplies a solid foundation, in a literal sense, for building new disciplines around the problem areas where existing disciplines intersect. As people and resources are centralized around problems, the resulting hybrid disciplines are stabilized and strengthened. Early on, the Center had, to some extent, been more of a harbor for researchers who were on the fringe of their departments within the university. Over time, it began supporting the growth of promising hybrids by fortifying them with new recruits from the outside. The Center's resources, facilities, and its reputation as an interdisciplinary harbor attracted desirable new faculty to the university. Sam was persuaded to bring his research program to the campus by the allure of the Center. The Biophysicist enticed a young rising star, "the most exciting molecular computation person that there was," to work with her by arranging for a Center appointment. The Center provided a "space" and an administrative mechanism for hiring scientists who did not fit in any of the academic units. The cognitive science group became more successful and active as a result of new hires who were recruited from outside the university. "It created a kind of center of gravity there, and all these other guys who had been sort of hanging back [in their departments] suddenly wanted to come." By enlisting researchers who were aligned with certain problem areas, the Center engaged in active hybrid-building, and the new recruits acted as trajection elements, providing the impetus for others from the established disciplinary unit to become involved in the newer emerging area.

Equipment also attracted people to the Center, and as researchers exploited the available instrumentation they reinforced research areas and approaches that depend on specific instrumentation. As we have seen, the sophisticated computational facilities encouraged theoretical work, sometimes at the expense of laboratory experimentation. Researchers and their students moved in directions that make use of the Center's technology. For example, the Animal Learning Specialist had branched off into large-scale database development, an area that would never have crossed his path or been adequately supported if he was working out of his regular academic department. Grant opportunities influence new research directions in concert with the technology. With more federal resources being earmarked for educational development, a couple of scientists had shifted part of their efforts to designing computer systems for education. The Device Physicist, who had no background in this type of work, claimed that computer instructional technology might end up being his "real claim to fame."

Internally allocated resources are invested and awarded for hybrid development, and certain research areas end up being advocated over others. Even at an institution devoted to interdisciplinary research, interdisciplinarity may not be the overriding consideration in such decisions. One researcher shared a story about a grant given in a recent graduate student competition. By his standards, the student doing the most exemplary interdisciplinary work was passed over. "She was looking at complex systems and human actions in a way that nobody else was doing, as far as I can tell. This was a real opportunity to pull these things together." He believed that the students who received the funding, while very bright and talented, were not as innovative, and clearly not as interdisciplinary. The probability of success may be more important than novelty in such funding decisions. As Tom explained in relation to garnering external grants, it is easier and safer to pursue research that is not too new, and this can apparently hold true within institutions designed to foster new areas of inquiry. Several other researchers pointed out that "successful" scientists would continue to have the strongest presence at the Center. Success is not an unreasonable criterion. However, it was not clear if the definition of success had been transformed to fit the demands of interdisciplinary work. Furthermore, no one commented on how long the lack of success would be tolerated before researchers would be sent back to their department.

The work climate at the Center is an important stabilizing factor. Center-based researchers coveted their work situation, stressing the importance of having a place and colleagues, rarely mentioning the state-of-the-art facilities. They believed they had the best possible arrangement, and for some the physical isolation of the Center was part of the attraction. They had an "ivory tower" away from the distractions of departmental business and politics. This distance also created a tension. The researchers preferred a sustained presence in the Center, but that was not always achievable. The Center-based researchers, who nearly all have full-time appointments with the institution, understood that they could not completely divorce themselves from their home departments—the tenure-granting bodies. Maintaining a positive presence in one's academic department is crucial, and those who are not full-time at the Center have learned that part-time appointments are rarely tenable.

> Unless you really pull up stakes and are over there 100%, you know it really makes the problem worse. . . . People can know you are part-time, but they can't reduce their expectations in half. They don't say, oh well John is doing 60% of what this full-time person is doing. All they think is John is doing less. (The Movement Specialist)

This conflict can be particularly difficult for new assistant professors; they need special mentoring to help them manipulate the social relations of the their interdisciplinary placement.

> We have learned from other people's unfortunate experiences. When a department hires somebody new who belongs over here for intellectual reasons, they have to warn the person that they shouldn't spend more than half of their time here. They should make themselves physically present [in the department] and make sure that they are socially and intellectually part of the life of that building. Otherwise, you know, people just don't have collegial feelings. (The Language Modeler)

In accordance with these complications, the profile of the Center membership had shifted over the first several years to encompass more full-time associates and a less diversified set of research groups than in the beginning. The original planners of the Center had expected that the members would spend part of their research time in their departments and part at the Center. Referred to by some as the "two office problem," having two research locations did not prove workable for many. According to the Organizer, in practice people ended up making binary decisions, either moving to the Center full-time or not at all. Some of the prospective research groups never moved in, and many of the part-time members dropped out. This had the effect of making some groups narrower because "a lot of the fringe type people disappeared." The population became more settled as the researchers at the margins returned to their departments and the Center concentrated on recruiting and retaining full-time members with track records of successful research.

COLLABORATION

Collaboration is a widely practiced interdisciplinary work approach that accommodates the needs of many kinds of researchers. Among the scientists studied, some worked primarily in a collaborative research mode, and all the researchers collaborated to some degree. The generalists placed more emphasis on their individual learning and research progress, but they depended on others to attain their goals and took part in collaborative projects on occasion. True collaborators worked more exclusively in groups and were invested in the group outcome. Of the many cooperative projects described by the scientists, there was considerable variation in the mix of intellectual domains and the level of integration across disciplines. The Organizer characterized integrative projects as "something that requires real doing back and forth on both sides." Additive collaborations, he explained, are those "where it is just a cookie cutter sort of thing."

> One of the things that makes for a collaboration is where work has to be done at both ends. That is, the theory isn't ready-made to solve this, to attack the data, and the experimental data at hand aren't precisely the data which the theoretician would ask for if he were going to test his theories.

Tom Tate was involved in several collaborative projects, including a couple that the Organizer might consider "cookie-cutter" in nature. One clearly was not. It covered a broad disciplinary range and involved a deep level of interaction, or "real doing back and forth." Tom is a versatile collaborator who understands how to implement different types of cooperative projects. In the collaborations he described, each combination of investigators created a unique problem-solving team. One of Tom's project teams was a highly effective mix that communicated and interacted well even though the researchers came from different disciplinary cultures.

> In mechanical engineering there is this guy Jack who has spent a good bit of time doing theory. Well, he doesn't do experiments, I do experiments. . . . And the more we talked the more it became quite clear that he knows how to analyze these data better than I do, and I know how to do the experiments quite reasonably. But there's this whole separate literature, and he knows that literature cold. I don't even have to touch that literature. I can say, is thus and thus? He says, oh yeah I remember thus and so was written in 1963 and I've got a reprint in my file and I will send you a copy. I just use him. I don't have to do anything else.

As an example of the Organizer's additive model of collaborative research, this project did not result in a high level of cross-domain knowledge integration. It was, however, the easiest to sustain in day-to-day practice. Tom continued:

> But you see this is where collaboration can work well if there's one person who has a huge expertise and a hole and the other person knows everything inside the hole but doesn't have the other expertise. Together you form a coherent unit and it works.

Previous experience working with different kinds of scientists is an obvious advantage. This particular engineer had worked with chemists in the past, and Tom was impressed with the interaction skills he had learned. "He can shift into my mindset and pull me over to his way of looking at things." The fluidity of this collaboration arose not only because of the complimentary combination of expertise, but also because the team members were both experienced at working and communicating with researchers from other backgrounds.

The next project Tom talked about was with a theoretician with a background in chemistry, not an outside discipline. They had both been trying to get funding for independent projects.

> So we were reading our reviews to each other over the phone. Her review said good theoretician, can't do an experiment. Mine said, decent experimentalist, doesn't know the theory. Oh yeah, we think we can cure this one. (Laughs).

Since they were both chemists the language did not detract from the ideas. The proposal was easily constructed, and they received the grant. In this case, combining a theoretical approach with Tom's experimental expertise resulted in a fundable project that turned out to be a harmonious and productive team effort. In Tom's experience, the serious barriers to communication lie between the disciplines not necessarily in the divide between experimentalists and theoreticians.

Tom had also recently written an interdisciplinary review article with the same collaborator. They developed the paper together, with each contributor bringing a different literature and perspective to the topic and working from distant sites exchanging revisions via e-mail. With a project like this, the most difficult collaborative effort goes into blending the divergent points of view. According to Tom, the paper achieved their aims and came about in "the way interdisciplinary work is supposed to happen." The final product was a comprehensive and integrative review that crossed the disciplines related to the topic, which shows that two problem-centered chemists can create an interdisciplinary product.

A collaboration with a theoretical physicist stands in stark contrast to Tom's other positive joint projects. It had been a challenging and frustrating long-term effort that seemed to be losing momentum. The written proposal for the project reflected the lack of convergence between him and the physicist—the document itself was an explicit example of the cultural barriers between disciplines. Two versions had been turned down for funding for what seemed like obvious reasons to Tom.

> You'd be reading along and I'd be talking about experiments, and then we'd get to a section where we needed his stuff and the language changed—the equations changed, the symbols changed, everything became dimensionless, and it was not even clear to me what he was talking about. It's not clear to me how to take my experimental data and put it into his theory and after his theory gets done crunching it—I know what it is supposed to do. It should give me a series of numbers to make use of to put out controls to my equipment.

The complications were not just in the writing and terminology; styles of thought were involved.

> The mind-sets, the ideas of what knobs you have to control and how you control them are so different that to put together a coherent description of that, otherwise known as a proposal, is extremely hard. . . . There is something that just doesn't really connect there.

Like the Language Modeler who spent hours deciphering usages of the word "gene," Tom was surprised at the incongruity of expression between scientific communities. I later found out that the project languished and the third proposal never happened.

Communication is the agent of collaboration. In boundary crossing work it cannot be assumed that the chosen words will convey the intended message. In relation to her studies of literature and science, Hayles suggests that scientists are trained to hold a "gift-wrap idea of language,"[3] that is, to see language as a package for handing an idea to another person. They do not consider what happens to the idea when it is converted into words or received by another person. Scholars trained in literature, she contends, learn that no two verbal statements can be the same, and that the wrapping counts as much as the idea. Tom and his physicist collaborator clearly had trouble unwrapping each other's ideas, and to the detriment of their funding possibilities, they never managed to repackage their contributions into one coherent and convincing text.

The scientists understood that collaborative work produces more than grant money and scientific results. Some generalists, for example, did not consider cooperative projects efficient because of the management effort involved, but they valued the process for its ability to broaden their perspectives and provide new insights. The intellectual exchange that results from having to work together leads to new ways of thinking about and solving problems. Unfortunately, such accomplishments are not evident in the tangible output of the completed project and go unrecognized in the reward system.

Collaboration is a highly adaptable research strategy. Every combination of researchers creates a different dynamic, and the relationships that form through collaborative projects and the activities that work well within various collaborative structures give us insights into how to construct workable research groups. Tom's three joint projects consisted of different combinations of scientific orientations and skills, but within each project the contributors' work and their interactions were directed at one specific research problem. The additive formula should not be discounted. Across the cases there were accounts of a number of such projects, and their success often did not depend on the extensive resources provided by the Center. For the more integrative initiatives, the Center played a larger role, and the translation work between languages and styles of thought was formidable.

THE IMPORT/EXPORT IMBALANCE

Set against the stasis that can be built by researchers and institutions through leeway, balance, centralization, and collaboration, the flow of information remains surprisingly unsteady in the interdisciplinary process. Unlike the dichotomies discussed above, the relationship between information import and export is askew. As became evident in Chapter 2, the act of importing information was a standard part of the research process for all the researchers, and different boundary crossing modes are associated with different types of import strategies. All the scientists gathered information from peripheral domains, relying on personal networks and conferences or employing a team approach. Many probed for information in new territories, and it was common for them to integrate a systematic learning program into their research process or utilize intermediaries for developing new areas of expertise. In comparison, information export, the *active* delivery of information across boundaries, was much less regular across cases. To review, only a few researchers worked at reaching multiple audiences, and those who had attempted to write for general audiences were not convinced that they had done so effectively. The researchers appreciated the "really good communicators," but their research aims did not necessarily include trying to be one. While researchers were highly aware of the language problems involved in importing information from outside disciplines, only a few consciously used language to communicate with their different constituencies.

The researchers were aware that I had performed a citation analysis of their publications to develop a profile of the range of their audience. While quite a few were surprised and pleased to learn that their work had penetrated other disciplines, most had not set out to export their results and ideas beyond their core research community. It is possible that the researchers did not have the leeway to support export practices, but they also did not acknowledge any potential benefits to active export. Two advantages seem obvious in relation to developing leeway and overcoming cultural constraints. Recognition in outside disciplines should contribute to symbolic leeway, leading to connections with peers and more freedom to expand into the periphery. Moreover, once a researcher has been acknowledged in a peripheral scientific community, the acculturation process should progress more easily.

Overall, the lack of equilibrium between import and export appears to be an accepted condition of interdisciplinary scientific research. Even at the Center, where the exchange of information across boundaries was a high priority, increasing export was not being stressed, except in terms of making public relations information more visible electronically through institutional web sites. At the time of the interviews, a few of the research groups had begun to make limited information about their projects available on the Internet, and the administration had started exploring ways of using web sites to showcase the Center's programs. Information sharing was not an explicit objective at the individual or organizational level.

In interdisciplinary research a high level of cooperation exists alongside strong competitiveness. Cooperation is often the most effective way to complete a specific project, but within and between fields there is intense competition for resources, authority, and territory. Rivalry contributes to the differential between import and export. Aggressive import helps individuals make progress on a problem and may lead to important new discoveries. Aggressive export, on the other hand, enriches adjacent domains and could lead to the advancement or encroachment of another discipline. According to Sam Smith, leading scientists in established disciplines have a "Darwinian urge to carry on their species." He explained that when scientists sense endangerment of their field they respond by they stealing opportunities for growth in new areas and eliminating competing research fields by outgrowing them.

> Worse, scientists of established disciplines are the gatekeepers for hiring, tenure decisions, academic honors, funding. In the Darwinian struggle of the disciplines, these scientists make use of their power. It would be malicious to state that this is done consciously. The scientists are deeply convinced that they do the best thing, but the outcome is disastrous for emerging disciplines.

Sam used a sporting metaphor to describe more calculated competitive tactics. Researchers, he explained, use a defensive soccer strategy; they kick the ball far from the goal, misdirecting the path of research in a problem area. "Scientists will kick it way over there if they know a guy has intentions to go here." Clearly, active export conflicts with the motives behind this type of maneuver. Other defenses are aimed at discrediting individuals rather than protecting critical areas for discovery.

> They will say that this scientist, of a new discipline, is obviously a bad scientist because according to our values he doesn't measure up. When you look at it, they have good values, but they are good only for a particular discipline and do not apply to the discipline of the scientist judged; yet they use it again and again.

Junior interdisciplinarians who attempt to export ideas into new territories where they have no allies or where values are still being negotiated are especially vulnerable to such attacks.

Lack of leeway and competition may continue to keep researchers from actively exporting their findings and ideas to other domains, with import remaining the emphasis of boundary crossing information work. There is a clear opportunity here for information science to assist in the cycle of accumulation by facilitating dissemination across disciplinary boundaries. To do so, information providers will need to set their goals beyond providing access and begin concentrating on how to promote interaction and synthesis within the research process. In the final chapter, I present a framework for thinking about the essential elements of interdisciplinary research and identify information initiatives for fostering interdisciplinary progress.

---

[1] This group includes the scientists who had a biophysics focus in training and those who were currently addressing biophysical research problems.

[2] This assessment takes into consideration the fact that, in general, those working in computer science and the behavioral science domains tended to publish less frequently than the scientists in all other areas.

[3] See the published e-mail conversation between Hayles and a practicing scientist (Stites 1993, 48) and the introduction to Hayles (1991).

Chapter 5.

CONCLUSION

ACCUMULATION AS RESEARCH WORK

The preceding chapters exposed and explicated the many elements of the accumulation process that are at play in the conduct of interdisciplinary research. Accumulation work continually mobilizes information resources to keep science moving, and it becomes more complex as researchers traject boundaries in the pursuit of new interdisciplinary knowledge. Much of the research process might appear unremarkable to the typical observer and perhaps even to people working on the inside of science. Administrators and technicians may get only a glimpse here and there of the system of individuals, materials, practices, and conditions involved in research work. In fact, much accumulation work is taken for granted by the scientists themselves, who as a rule do not reflect on this part of their job until they run into problems.

Interdisciplinary accumulation is complicated by the need to bring together knowledge from multiple domains. Structural and cultural boundaries get in the way. To successfully promote interdisciplinary inquiry, all aspects of the accumulation process need to be recognized and supported, especially the work at the boundaries between scientific communities. Thus my analysis has concentrated on the boundary crossing practices and strategies of the researchers, the major influences in the work setting, and how the interdisciplinary research process is currently accommodated by formal and informal bridge-building structures. There is one final question to address: What have we learned that can help us build better environments for interdisciplinary research?

We have observed that the path of problem solving is not singular, but manifold and complex. It is not fixed, but variable. There are many ways of doing science, and approaches accepted by one community may be rebuffed by another. Consider the practice of collaboration. It is the dominant strategy for managing interdisciplinary research in many sciences, but in certain domains it is discouraged and devalued. For example, the norms in the Language Modeler's dual research worlds diverged on this issue. At the

Center, collaboration was not only valued but also expected. The culture was different in his home department: "I've served on search committees and promotion and tenure committees and seen what happens to people whose work is collaborative . . . people say the other guy did it all." This type of incongruence between fields limits collaborative diversity. In the Language Modelers area of study, identifying a linguist who is free to do teamwork on problems in language production is likely to be much more difficult than finding a psychologist, physiologist, or a computer scientist.

It is not the act of collaboration in itself, however, that solves a research problem. Some collaborations work smoothly, some are painstaking, and some produce a lot of learning but no visible results. Progress takes place as group members interact and perform work in spite of the boundaries that exist between them. It may be commonplace for a chemist to collaborate with a physicist. As a chemist, however, Tom Tate's strategy of training and deploying a translator as part of a collaborative team was not standard practice, and the long process of give-and-take as the group inched toward experimental and theoretical convergence could easily go unnoticed. The Movement Specialist spent much of his time verbally interacting with people in industry and government instead of writing because that was how he "learned the language, metaphors, and concrete facts" that allowed him to make progress. This work may not result in products that are obvious to other members of the academic world. Nonetheless, he was acting as a cultural broker, making important advances in cross-cultural communication and understanding.

Incremental steps and activities that may appear inconsequential are the building blocks of the broad connections that traject disciplines. Poorly attended and seemingly unsuccessful general presentations can have a significant impact on theory development, albeit for a select number of individuals in attendance. Reading and interacting far afield can sustain or even transform a career while seeming frivolous to colleagues. Problem-centered research reviews are critical for training newcomers to a field and for documenting the reconfiguration of knowledge domains, but do not count for much in the present academic reward system. Intangibles also play an important role in boundary crossing research. Unexpected stimuli and aesthetic thought processes, such as "pattern recognition," help in the construction of conceptual linkages.

For the researchers in this study, the most effective platforms for these low-visibility research activities are those that allow active interchange. The projects, meetings, and personal networks that bring people together from different backgrounds provide a scaffold where real progress can be made toward what often appear to be the intractable problems of cross-cultural communication and overload. As Neil Nash pointed out, in his field it took years of explication between the "physical types" and the biologists before meaningful communication began to take place. These interfaces are also where the mass of information accumulated by the researchers begins to take shape. Through interactions with others who have complementary expertise, researchers weigh and sort the information they encounter and collect to determine the most relevant for the problems under investigation.

INVISIBLE CONSITUENCIES

One might expect the cultural barriers between disciplines to make informal communication a less vital part of scientific research. Instead, personal contacts remain as central in interdisciplinary science as they are in disciplinary science. Colleague networks continue to rear the research establishment, primarily because they are the most effective mechanism for personalized, research related exchange. The informal information networks in interdisciplinary problem areas are not the same as the invisible colleges of disciplinary science, however. When Price (1961) introduced the idea of the invisible college, he was making reference to the inner circle in a scientific specialty, the active scientists who meet at select conferences, collaborate on projects, travel between each other's research sites, and circulate research reports to one another. This type of "ingroup," as Price called it, tends to be made up of productive, high-status scientists who act as information dissemination nodes, fostering development of the field within a self-reinforcing social circle. By extension, ingroups also control who gains prestige and the fate of new ideas in a field (Crawford 1971, Griffith, Jahn, and Miller 1971, Crane 1972, Cronin 1982).

We have seen that there are powerful disciplinary invisible colleges that the researchers must contend with that act as elite power groups and uphold entrenched standards of conduct. These are not the informal communication networks that sustain boundary crossing inquiry. Working interdisciplinary networks are better characterized as constituencies than colleges. Invisible constituencies form around interdisciplinary scientific problems. The networks are more open, loosely organized, heterogeneous, and distributed than invisible colleges, and network associates tend to function more as advocates than critics. Interdisciplinary areas of research have not generally matured to the point that prominent experts can regulate them. Moreover, a hierarchical system of control would not be effective for managing the flow of information, ideas, and influences across boundaries.

Networked information systems have made it easier for researchers to communicate and gather information informally. Individuals can more readily identify kindred scientists and laboratories through the Internet, send queries and establish contact with potential colleagues via e-mail, and locate fresh research results in electronic preprint archives.[1] As hybrid research areas become stabilized within institutions, and as highly influential figures emerge in these fields, loose constituencies will develop into more stable collegelike social networks. At the same time, the fluidity of electronic communication should permanently ease researchers' ability to make contact with other scientists, collect relevant information across boundaries, and build more egalitarian communities of researchers. These constituencies will become visible over time through the connections that grow between research groups represented on the Internet, perhaps subverting some of the authority held by invisible colleges.

INTERACTION AND LEEWAY

The picture of research work that grew out of the interview data is multidimensional and mutable, yet there were firm patterns that held across the cases. All the scientists, regardless of their preferred research mode, used personal networks for gathering information and preferred problem-centered work associations over more disciplinary structures. In addition, all the researchers faced challenges in two distinct areas. Cross-cultural complications and information overload interfered with the accumulation process of team leaders, collaborators, and generalists. On the other hand, the group of researchers, representing a range of biological, physical, behavioral, and computational scientists, exhibited considerable variation in their level of interaction with other researchers. On an interaction continuum, as illustrated in Figure 4, the highly participatory researchers—strict collaborators and team leaders—are at one end, and the individualistic generalists are at the other end. The researchers highlighted in the last chapter, who take a more problem oriented, multimodal approach, lie in between the poles of the highly participatory and individualistic researchers, essentially adding a forth mode to the original model proposed in Chapter 1.

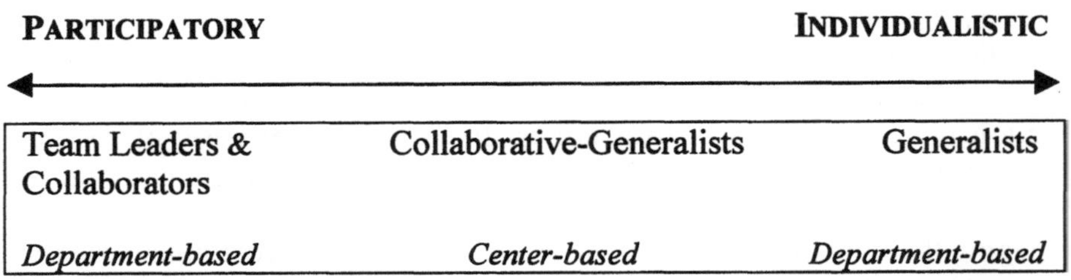

Figure 4. Research interaction continuum

It is informative to consider the researchers' interaction tendencies in relation to the location of their research operations. The researchers in the middle of the continuum had the strongest ties to the Center. All but one of the twelve scientists who housed their research projects at the Center were moderately interactive; they consulted frequently with people in other disciplines and participated in research projects with scientists in other fields. This count includes all of the collaborative generalists—the psychologists and biophysicists profiled in the last chapter—and includes researchers from linguistics and biology. The remaining Center-based researcher is a quintessential team leader with few generalist tendencies. In contrast, those who maintained their research base in their academic department tended to be at either end of the interaction continuum. The highly interactive researchers who engage almost exclusively in collaborative work and the other team leaders who manage large research groups are at the participatory end. The generalists who concentrate their own knowledge base development and prefer to work independently lie at the other end. The Center is not the locus of work for either of these groups; instead they rely on shadow structures to frame their cross-disciplinary connections, and they mix more with faculty in other units in the university and with scientists in government and industry than with Center affiliates.

All the approaches along the continuum are important ways of making progress on scientific problems. The various approaches are also dependent on highly specialized research that does not cross disciplines, for it often provides the substance that is accumulated and reconfigured in interdisciplinary research. In terms of boundary crossing effectiveness, the scientists who combined participatory and individualistic strategies had notable success managing to be both productive and integrative and offer the best model of interdisciplinary practice. They apply a broad repertoire of boundary crossing techniques and as a result are best equipped to manage the dichotomies involved in the interdisciplinary research process. These problem oriented researchers alter their level of dependence on the knowledge of other scientists to balance focus and breadth and to coordinate accumulation activities.

It follows that support systems for interdisciplinary research need to be designed to sustain both participatory and individual work approaches. A comprehensive infrastructure should take into consideration the fact that problem-centered research does not fit neatly into any one framework of inquiry. Solutions are woven together from multiple knowledge domains by researchers who not only vary in their work styles and preferences, but who may change their approach depending on the problem being investigated. My application of the concept of leeway reflects the need for flexibility in research environments that wish to promote boundary crossing inquiry. Adequate levels of leeway, in the form of articulation work, intellectual proximity, and technological and symbolic assets, clear the way for researchers to address problems from a variety of contexts.

In practice the participatory orientation requires less leeway than the generalist approach. Most of the highly participatory researchers have achieved their goals without relying heavily on leeway-producing mechanisms provided by the Center. They have developed work strategies that do not require an additional overlay of space, resources, and articulation work. While participatory researchers do benefit from the Center's symbolic leeway, they engineer their own intellectual proximity through local collaborative projects and personal colleague networks. Researchers who adopt an individualistic approach need more leeway because of their emphasis on information probing and learning. Thus we can add another dimension to the interaction continuum, as illustrated in Figure 5. Researchers on the participatory end require fewer types of leeway, and the leeway conditions must increase for those further along the continuum. Individualistic researchers require the highest level of leeway. Symbolic leeway is particularly critical for them since in most academic situations intellectual products are explicitly rewarded while there are few incentives to undertake the laborious processes of knowledge base development. Administrative sanctioning of both participatory and generalist orientations is one important step toward symbolic leeway.

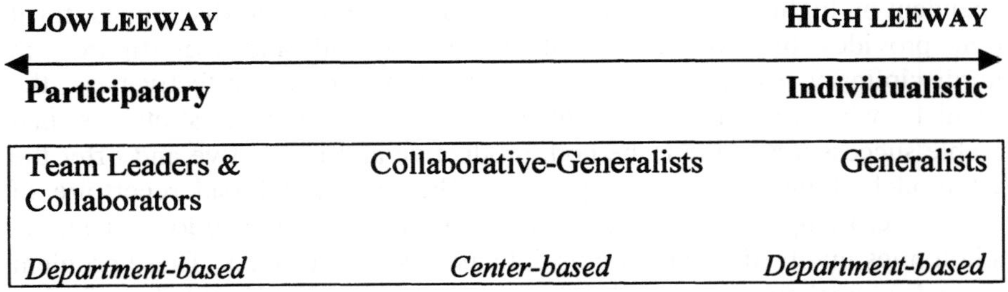

Figure 5. Research interaction continuum in relation to leeway

For many of the scientists, the Center's facilities and atmosphere are almost ideal. As such, the Center serves as an instructive guide for designing and readapting our research organizations to meet the requirements of interdisciplinary inquiry. We can learn much about boundary crossing research from the activities, experiences, and perceptions of the scientists at the Center, especially if we look beyond the factors that are dependent on a resource-rich environment. Throughout my analysis I have stressed the importance of understanding how boundary crossing research can be accomplished in more typical research organizations. The researchers at the participatory end of the continuum who were not based at the Center represent the most attainable form of interdisciplinarity. They have not required extensive new resources, and they excel at managing information accumulation and overload. The most significant element of the participatory approaches has to do with the outcomes of group work.

COPRESENCE AND ACCULTURATION

Working multidisciplinary groups are effective centers for producing research but also for learning through what Brown and Duguid (1991) refer to as "copresence." Judging from Tom's experiences with many different collaborative combinations, copresence that spans large disciplinary divides is the most lucrative for learning. Laboratory environments that house researchers with different backgrounds are a microcosm for rich, boundary crossing exchange. Sam worked hard to orchestrate a diverse copresence in his lab that could fill the teaching gaps in the university. Learning at the "periphery of practice" contextualizes knowledge, gives it meaning, and makes it easier to assimilate (Brown and Duguid 1991, Lave and Wenger 1991). It other words, it was probably more effective for Sam's "little Einsteins" to learn about biological computation by rubbing shoulders with computer scientists and biologists in the lab than it would have been for them to enroll in more general course work. The lab allowed more extensive social interaction where the viewpoints and vocabulary of scientific communities could be exchanged along with content knowledge and know-how, with acculturation happening as part of the learning process.

Learning through copresence in a laboratory or collaborative situation is especially well suited to interdisciplinary work because the exchange of information is centered on a specific project or a distinct problem area. Problem solving becomes a collective aim. Extraneous or marginally related material can be filtered out by the multiple expertise, decreasing overload complications thus making it possible to do what would otherwise be impossible (Wilson 1996). Departmental multidisciplinary copresence does not have the same impact. The Movement Specialist and the Computational Designer experienced alienation in their academic units where faculty from different backgrounds where centralized but were not working together on similar problems. Multidisciplinary copresence within a very interactive department could conceivably provide an enriched learning environment primed for new discoveries. The "interplay of separate communities with independent world views" is important for innovation (Brown and Duguid 1991, 54), and under some circumstances groups with competing interests may participate in productive interaction (von Hippel 1988). Nevertheless, the best examples of innovative achievement revealed in this study were generated within or across diverse laboratory teams, as in the case of the Photosynthesis Specialist's in-residence program with another lab. His team produced new methods, improved operations for both labs, and produced experimental results that had implications for the field more generally.

Both participatory and individualistic interdisciplinarians are faced with intimidating acculturation burdens. Problems surface for generalists as they work at becoming members of peripheral communities. Highly collaborative researchers face different complications since they may not attempt to actually gain entrance into new groups. Their interactions are more project dependent. Some collaborations do not constitute substantive copresense or result in deep knowledge development or lucrative exchange of information across cultural divides. However, participants in a collaboration must develop a common base of understanding, and intermediaries who cross boundaries and return to their home territory to work or translate must be able to function in multiple cultures. When the AI Specialist struggled to negotiate the necessary degree of complexity for talking and working with physical scientists, he was combating superficiality and working toward real cross-cultural comprehension. When he sent an emissary to live and learn in another culture, he arranged an effective short-term copresence situation through the delegation of presence. Delegation, according to Latour (1992), is the transformation of major work into a more minor effort performed by either humans or machines. The delegation to intermediaries, observed in the AI specialist's projects and in the exchange between the photosynthesis labs, could be considered more minor than other components of the projects because it was work performed by students, but the tasks were by no means insignificant in the overall equation for successful innovation. The judicious delegation of translation work and finding ways to turn it into articulation work will do much to advance interdisciplinary initiatives. But, regardless of shifts in work responsibilities, some research team members will need to be adept at both scientific and interdisciplinary communication.

While group work is a defining feature of participatory research, it is only one of many strategies used by the collaborative generalists. Instead of being the modus operandi, group work is employed when it is the best way to solve the problem, to distribute labor, or to sustain a research core while building new bases of knowledge for other projects. Problem-centered researchers, working independently or as part of a team, need to be able to understand, evaluate, and apply knowledge from multiple scientific domains. To accomplish this they enlist people, objects, methods, and information and develop skills that help them interact outside their core intellectual community. They learn to build networks and exchange information across boundaries and across cultures. For interdisciplinary progress, information exchanges and knowledge-building activities must be broad and focused, and accumulation must occur at the core and periphery. Copresence is the most effective mechanism for information exchange and contextual learning. It is a context where participatory and individualistic practices can blend to balance the dichotomies of the research process.

MANAGING MULTIPLICITY

Supporting interdisciplinary work at the organizational level means supporting the many ways of doing science. As the Photosynthesis Specialist pointed out, "there is a need for people of many different abilities and temperaments in science." Academic departments are the most typical administrative structure for organizing science in universities, and these units tend to be tightly aligned with long-established disciplines. Over the past few decades the archaic nature of the university structure has been debated and alternatives to the system have been proposed (Perkins 1973, Alpert 1985), but the standard is not likely to change much in the near future. Boundary crossing researchers will need to coexist with disciplinary researchers, and continued interaction between the two perspectives is essential and workable within an accommodating research environment.

Discipline-based organizations can offer symbolic leeway for boundary crossing research by instituting flexible expectations for career trajectories. Researchers will progress differently depending on their research orientation and how much of their work takes place in core and peripheral domains. An accurate calculation of interdisciplinary achievement needs to factor in acculturation and translation work. Staggered leaps of learning, interacting, and production should be considered the norm, especially for generalists, and all boundary crossers are likely to experience periods that are devoted to working toward coherence. As was the case with one of Tom's ambitious integrative projects, getting past the proposal stage can be a significant achievement since writing a cohesive document takes iterative translation work. And, sometimes the best efforts fail.

Departmental "black holes" suppress variant styles, skills, and perspectives, and individuals who attempt to move away from those gravitational forces are taking risks with their careers. Researchers who attempt to work outside of disciplinary and departmental boundaries face internal, as well as external, alienation. They may experience rejection by their local communities, as described by the Movement Specialist who was an outlier in his department; or they may be dismissed by the larger scientific community to which they are trying to contribute, like the Neurophysiologist. Moreover, attaining membership at a separate center for interdisciplinary research is a special distinction that can have mixed outcomes. While it can increase prestige and acceptance in cognate research areas, it may also lead to estrangement from established local constituencies.

Frameworks other than departments and special centers help to marshal and foster boundary crossing research. As we have seen, informal shadow structures support many local coalitions. Researchers who believe that their disciplines are insufficient or obsolete, such as Sam and the Computational Designer, search for new ways to solve problems and for new colleagues with the expertise to help them. In doing so, they make links that form emergent groups. On the local level, these groups can be as large as the combined faculty of two or more departments or as small as a cross-departmental partnership between two individuals. Once identified, these clusters can be fueled without disrupting their informal, distributed structure. For example, while the cognitive scientists welcomed their move to the Center, other steps might have stabilized the ensemble. Congregations can be encouraged with recognition rather than formally instituted resources. In fact, some attributes of the Center worked against constructive reconfiguration, as with Neil's biological students who were lured into becoming strict computationalists through exposure to high-level technologies. Furthermore, while centralization increased opportunities for interaction between the disciplines on a daily basis, discipline-building efforts worked to ensconce the membership at the Center. Physical proximity promoted more casual cross-disciplinary interaction, but as the department-based cases in this study demonstrated, it was not essential for carrying out research projects. Physical centralization is a necessary anchor, however, if an organization aims to go beyond facilitating innovative research to constructing new hybrid disciplines.

Individuals make a significant contribution to the overall process of interdisciplinary integration through their multiple memberships. Establishing and sustaining memberships—disciplinary or interdisciplinary, formal or informal—is a management problem. Some of the scientists had many local affiliations, such as the Biophysicist who was active within a number of groups at the Center and with several other formal and emergent groups across campus. Neil's membership profile was also complex, but much less localized. He had strong ties to national and international problem-based groups and a much weaker association with the Center. As a whole, between their departmental and Center assignments and their other affiliations, the researchers' positions add up to much more than what would be considered a full-time appointment. Active interdisciplinary researchers may, in fact, rarely be able to perform as a full member of any one group. Although it is unwieldy personally and administratively, juggling many partial commitments should be considered standard career progress for interdisciplinary researchers. Devotion to any single unit is an unrealistic and unproductive expectation.

Maintaining multiple memberships is complicated, but the process of gaining entrance into these foreign communities is no less formidable. Where an individual might tend to be receptive to new perspectives, a group can function as a very closed system. It took Tom more than ten years to break into one new community, and Sam hated the process with "every molecule" of his being. Getting inside requires seeking out a community and usually participating in it over an extensive period of time. It seems, however, that short, concentrated efforts may be effective for some degree of initiation. Several researchers considered intensive workshops to be one of their most important accumulation activities. These working meetings functioned as acculturation sabbaticals where work gets done and learning take place completely within the context of the outside culture, in a concentrated copresence situation.

Cross-cultural communication requires comprehension between disciplines. The capacity for comprehension evolves over the course of exchange or discourse between communities, which Klein (1996) refers to as the "rhetorical, social, and political negotiation" or communicative action of interdisciplinarity (221). Communication occurs in many settings—in labs, classrooms, at meetings, via e-mail, and on the pages of books and journals. As we have seen, successful boundary crossing conversations require a lot of explanation between participants. If dialectics is the true method of interdisciplinary work, as Klein (1996) suggests, then multiple membership is the key to dialectic progress. The process of interdisciplinary translation and understanding occurs incrementally between diverse individuals and groups and work best in forums that foster questioning and explaining. Integration takes place as individuals develop multiple affiliations and reconfigure into new combinations where further translation takes place. As cross-membership increases, so will the transfer of language, culture, and knowledge across communities.

ACCUMULATION AS AN INFORMATION PROBLEM

Information is at the heart of the cycle of accumulation. Knowledge is gained through networks of people, objects, and actions, and information is a large part of what is accumulated through the interactions and negotiations within research networks. Moreover, information resources and information technologies act as catalysts for the accumulation of knowledge. With information at the center of both the content and the manner of accumulation, information science and information professionals are uniquely positioned to advance the interdisciplinary accumulation process. Since understanding boundary crossing work practices, accommodating variant research orientations, and supporting multiple memberships are central to fostering interdisciplinary research, information environments for interdisciplinary inquiry should embody the same attributes. Information systems and services for researchers need to provide access to our many heterogeneous bodies of knowledge, support the divergent approaches to problem solving applied by researchers, and allow communities of researchers to share their information resources and products.

Some of the scientists at the Center talked about new or improved information services that would be useful to them. In general, they described services to provide customized gathering of information across boundaries and assistance in understanding work from other scientific communities. In other words, they recognized a need for high-level articulation work. They identified a role for an additional player in the cycle of accumulation—a specialist to navigate, filter, and compile information, or serve as an "intelligent mediator."[2] As might be expected, these scientists wanted assistance finding, comprehending, and learning subject areas outside their expertise. The following excerpts are representative of the information needs expressed by the researchers:

> I would like to have some way of finding out what in the literature is the most important in the subjects I am interested in. An index of some sort that crossed Psychology, Computer Science, Linguistics, etc. would be nice.

> I need to identify a small percentage, some suitable subset of research in subject areas related to my work.

> [I need] an easy way to learn the basics about other disciplines without being an expert, an easy way to read about things.

> It would be helpful to be able to identify who is citing what, or what is being used or cited strongly, like what linguistic work is being heavily used by cognitive scientists.

Mediating Languages and Literatures

A few of the researchers suggested that the dissemination and export of information across scientific domains could be improved through efforts in information science, especially if the field were to address the problem of translating jargon into meaningful terminology for cognate research areas. Progress on vocabulary problems has been made on several important fronts. Techniques have been developed for automatically identifying sublanguages within disciplines (Haas 1997). This work has direct implications for improved domain-specific information retrieval, and it could also lead to a better understanding of the relationship between families of sublanguages and how language reflects the attributes of established and emerging fields. Other research aimed at enhancing information sharing and retrieval between subfields in molecular biology has made progress reconciling overlapping vocabularies by merging thesauri using a "concept space" approach to automatic thesaurus generation, (Chen et al. 1995, 1997). These studies have shown how complicated the language problems are across relatively narrow disciplinary divides. Information science has not yet adequately addressed the monumental socio-technical factors that involve the cultures and practices of people within research communities. Longitudinal tracking and detailed study of both the technical and social aspects of the creolization process may be necessary before more divergent vocabularies can be successfully cross-indexed or conjoined. In a sense, cross-disciplinary thesauri that break down language barriers constitute concrete representations of what might otherwise exist only as metaphorical communities (Pahre 1996)—domains that emerge around ideas and concepts but do not form true communities of practice. At some disciplinary intersections real intellectual communities may not be able to evolve until mechanisms for translation are in place.

In tandem with the development of information systems, information specialists are faced with the challenges of supporting the work of boundary crossing researchers in academic and industry settings. Librarians must be adept at searching both broadly and specifically across disciplines for problem-based subsets of information. They must be skilled at locating and selecting what one researcher referred to as an applicable "smattering" of relevant material. Developing these abilities takes more than training in information searching and retrieval. Information service providers need to understand the dynamics of knowledge in terms of domains and their content, as well as the context and conduct of research in practice. Highly skilled searchers are continually hindered by the way our libraries and information systems store, manage, and disseminate the scholarly literature.[3] The published materials that relate to a given subject are unpredictable, and the tools for identifying and locating them are uncoupled and far from comprehensive in their coverage. To illustrate the scattered nature of literatures, Mann (1993) offers a partial list of 30 standard bibliographic indexes that contribute to religious discourse. These include resources covering statistics, science and technology, business, art, and law. The tools are not the only limitation in the knowledge systems, however. Researchers' conceptual constructs of what is possible are also a problem. Information seekers are constrained by their own expectations of what exists and where it is likely to be located (Mann 1993), and the situation has become more complex with the growth of electronic information networks that add "unpublished" texts, pieces of documents, and endless ephemera to the mix of literatures.

Discovery and Synthesis

The cognitive paradox makes it impossible for people to fully exploit information storage and retrieval systems (Neisser 1976, de Mey 1982, Cole 1994). How can a person search for something that is unknown to them? In fact, by some definitions, a sensory experience may only qualify as "information" if it involves the perception of something completely new or modifies an existing knowledge structure (Neisser 1976). "Information has to be surprising or it is not information but just a record of what we already know" (Boulding 1983). In this sense, it is rare to encounter "true information" that has not been predicted or anticipated (Cole 1994). In addition to being rare, new information is difficult to understand since the receiver may have no relevant experiential framework for something novel and vocabulary exchange problems come into play. Yet scholars seem to value novelty above all else in their quests for information (Sandstrom 1994). As demonstrated by the probing activities of the Center researchers, the unknown and unfamiliar can have an important impact on the process of inquiry, moving it forward or into a new fruitful direction. Information systems, unfortunately, have not been designed to meet unknown or undetermined needs (Davies 1989). A model of scholarly information gathering proposed by Sandstrom (1994) applies optimal foraging theory to the information search process, equating scientists and scholars to hunter-gatherers and animal foragers, who learn to forage in "patchy information environments" to gather sufficient resources. Not just any provisions will sustain researchers, however. They thrive by being able to compile unique combinations of resources, and interdisciplinary researchers work within information environments that are especially patchy, making it difficult for them to bring together a body of material that is unique and relevant.

The new information that can be culled from overlapping intellectual domains is likely to be the most vital for interdisciplinary synthesis. Therefore systems promoting interdisciplinary innovation need to provide access to strategic interdisciplinary intersections and support the identification of subsets of novel material within them. While not nearly flexible enough, current database systems can be manipulated to promote discovery for synthesis. Certain techniques can be used to determine the potentially lucrative information spaces where subjects, concepts, and problems intersect or overlap. Searching techniques that favor recall over precision[4] are considered the most effective for retrieving overlap across subject areas and for identifying unfamiliar materials. In databases that are not narrowly specialized, search statements that omit terms specifying an area of application can retrieve relevant information from distant fields (Davies 1989). Progress has also been made in devising methods for identifying complementary combinations of information that lie within our vast stores of recorded knowledge. Swanson's (1986a, 1986b, 1987, 1988) studies are of particular interest. His work linking disconnected biomedical literatures suggests that there are countless important connections waiting to be uncovered and that text processing techniques can identify the most promising areas for synthesis.

The information gathering methods for "creating interdisciplinarity" formulated by White (1996) produce maximum license for making creative connections by focusing retrieval on the "set of specific ideas on what may be synthesized" (253). The creative process involved in synthesis may be enhanced through information technology, but, as White notes, creativity can't be completely reduced to algorithms. Noting a similar connection between creativity and synthesis, Bawden (1986) suggests information conditions to aid in the creative process. He recommends an information-rich environment with technologies that emphasize browsing, provide access to peripheral material, and represent analogies, patterns, and exceptions. The information practices of researchers at the Center confirm that all of these attributes would enhance boundary crossing inquiry. The interdisciplinary research process strives for creativity in problem solving. Moreover, the ability to creatively manipulate vast amounts of information from multiple domains has great potential for leading to new discoveries.

Retrieval tools are but one layer of the complex of stored information and knowledge. Documents make up much of the mass, and specific documents can act as interdisciplinary accumulation tools in their own right. Integrative literature has been recognized for some time as a critical resource for promoting synthesis across subject areas (Garvey 1979). In particular, integrative reviews of either a bibliographic or substantive nature (McInnis 1996) cut across subject domains to correlate separate pieces of related evidence (Davies 1989), and they can reveal the current state of progress in an area of study (Sandstrom 1994). They are packages of collected and partially synthesized cross-domain information that supplant extensive and difficult information gathering. Meta-analytic reviews (Becker 1996) are perhaps the most valuable since their aim is to produce an exhaustive compilation of relevant results, and they may include hard to locate fugitive literature (Rosenthal 1994, Rosenthal 1995).

Information specialists can exploit existing reviews to assist in information retrieval for cross-domain synthesis, but there are also compelling reasons for them to be involved in producing reviews. First, they have the knowledge, training, and experience to conduct comprehensive literature searches to find all material regardless of its origin (Smith 1996). As mentioned above, breadth of recall is more important than limited selection for multisubject retrieval. As Oxman and Guyatt (1993) assert, experts in a field may overemphasize personally salient works, overlooking small or less prominent studies. Reviews compiled by experts provide an important filtering function for a certain audience, but they may not lay the best groundwork for identifying new connections. Reviews for interdisciplinary synthesis need to allow readers to determine overlap and novelty based on their particular research questions and approaches. Second, information professionals face fewer deterrents to this kind of research work than academic researchers. Even though literature reviews are important information resources for scientists, there are few incentives for them to undertake this type of project. Tom Tate was committed to writing integrative reviews, but he also understood that successful scientists focus on conducting experiments and developing theories.

Journals devoted to the dissemination of review literature are unique resources with a high density of potential cross-domain connections. These publications are beginning to fill a long-standing need in interdisciplinary and multidisciplinary research areas for broad coverage that goes beyond current awareness functions (Shephard 1983, Davies 1989). As more review literature becomes available online,[5] it will be easier to trace and "mine" for new associations within and across bodies of literature.

CREATING INFORMATION LEEWAY

The principles of user-centered information services dictate that the development of information resources and systems should be aligned with researchers' practices and needs,[6] yet the practice of boundary crossing, a formidable part of innovative and integrative research, remains largely unsupported. The essence of interdisciplinary accumulation lies in the dialectic work accomplished as researchers traject boundaries and form new associations based on research problems. Our current systems of information, however, do not explicitly represent working aggregations of researchers and the intellectual associations that link them. Before we can improve information systems and services for boundary crossing research, we need to establish ways to identify cogent configurations of researchers and to understand their practices and needs in accomplishing research work within those problem-based structures.

Working Groups

We know that physics, chemistry, and psychology are not particularly meaningful knowledge domains for the researchers at the Center. We have also seen the insignificance of many of the official categories that are applied to the researchers throughout their careers. Ph.D., discipline, and academic department were not valid or complete descriptors for the scientists, and the Center's designations were just one of many research oriented affiliations in each scientist's constellation of memberships. The problems that scientists work on, however, were germane. Problem areas represent authentic and purposeful domains for organizing both research activities and the information systems that support those activities. Groups emerge around problem areas, as researchers associate and communicate in local networks, through e-mail across the Internet, and at seminars and conferences. The most well-grounded research groups form physically and virtually in many informal and unrecognized pockets of academe and in the professional world outside the university setting.

This study validated that one of the greatest challenges for information system development is the situational nature of information needs. While we can strive to create adaptable systems that can be customized to personal profiles, it is more realistic and more productive for dialectic purposes to focus on aggregations based on information use patterns—that is, groups of users from different subject backgrounds who address similar problem areas and have common information practices and needs. Within problem areas the interaction continuum portrays an important dimension of information behavior. Degrees of interaction are linked to the location and management of projects and personnel. The three modes—team leader, collaborator, and generalist—differ by level of interaction and are distinguished by information and knowledge development practice. Interaction requirements in conjunction with the boundary crossing modes indicate the levels of leeway needed for interdisciplinary work to proceed.

For interdisciplinary initiatives to thrive, resources need to be directed to relevant working groups. Detection and support of emergent communities in the workplace can foster working, learning, and innovation (Brown and Duguid 1991). Research is a unique type of work (Star 1989, Pickering 1992), and learning and exploration lay the foundation for the innovation and integration produced through interdisciplinary research work. Interdisciplinary researchers cross boundaries to explore and mobilize existing knowledge in new ways and to develop new perspectives on scientific problems. The formal and informal structures that support boundary crossing work, learning, and discovery provide important guides for organizing information and information services. While the number of local, national, and international emergent communities is too great to address each as a separate user group, these units can be combined into larger overlapping problem areas.

Mapping of problem areas can be done locally or globally through bibliometric analyses to identify subject intersections and to guide how scattered resources can be networked to enhance the ability to make important connections. Bibliometrics is suited to detecting the formation of synchronic subject relationships, and, if applied over time, diachronic configurations as well (McInnis 1996). In addition to bibliographic indicators, conference-based and correspondence based-networks provide the basis for more timely analyses of relationships between researchers. Once the links between areas of discourse are mapped, it becomes feasible to construct up-to-date frameworks for research services. In terms of organizing information resources, frameworks could take the form of "spectra" of literature that reflect the actual subject orientations of scholars as they change over time. White (1996) uses the term spectra to describe the results of a type of interdisciplinary literature search, but the concept also captures the cross section of scattered but related literatures that could readily be linked in networked collections.

Certain genres of literature can serve as organizational outlines for ordering spectra of literature around interdisciplinary topics. Digitized documents that include links to the full-text of each reference perform important articulation work that creates information leeway for researchers. Highly interdisciplinary linked documents do significant work by providing coverage of interrelated intellectual territories and access to otherwise disparate materials. As Tom indicated, research reviews in particular save "generation after generation from hitting their heads against the same problem." Reviews place the problem at the center, and the links between references and the source text create a web of information that spreads out from the problem, much like the process of footnote chasing. The structure is more reflective of the way knowledge spreads, connects, and grows than standard bibliographic tools. Unfortunately, as digital content becomes widely available, many directories and organization schemes continue to differentiate disciplines first and then attempt to fit hybrid information into these ill-suited categories.

The organization of information within cross-disciplinary spectra would be conducive to an interdisciplinary approach to research and need not preclude effective access for disciplinary research. In fact, at the local level, it is crucial that research libraries and information centers continue to support the conventional departments and established research institutes that feed the unofficial emergent research areas. This is part of the institutional problem of managing multiplicity. Information specialists will need to find ways to establish liaisons with the hybrid units of research as they materialize. Fortunately, recognizing and addressing the important role of learning in interdisciplinary research will enhance rather than detract from the pedagogical mission of academe. While information services uphold official departmental teaching efforts, the shadow curriculum that mirrors the emergent research groups needs attention as well. Students and faculty are involved in important supplementary programs of education, as was the case with Sam Smith, who was single-handedly attempting to fill the teaching gaps for students in his lab. To fully exploit situated learning, information services need to be *in* the labs, where researchers and students ask questions, pursue ideas, and conceive of and embark on projects.

Since scientific communities are made up of researchers who represent various points on the interaction continuum, it is necessary to accommodate the full range of information activities within a problem area. Two members of the complex systems group, a community that formed independently of the Center, serve as a good illustration of the divergent information practices used by scientists with the same problem interests. Tom, who was studying oscillating reactions, favored a participatory approach. In addition to active consultation with others, he scanned expansive literatures, relied on online literature searching for gathering information, and regularly attended certain specialized conferences. The Molecular Network Specialist, whose orientation was much further toward the generalist end of the continuum, tunneled through the literature following leads in articles, read general multidisciplinary science periodicals, scanned articles, reviewed textbooks, and preferred seminars over conferences.

This diagnosis of information practices has clear implications for system design. A workable system for complex systems research would need to account for Tom's approach by allowing access to all the chemistry, biology, and physics material relevant to complex systems. In terms of functions, the system would need to support general browsing of that material, as well as searching for specific keywords and subjects. The molecular networker was interested in principles of organization on a grander scale, ranging from machine design to the dynamics of the human organism. For his purposes, the content of an information system would need to include computer science, engineering, physiology, and psychology, in addition to chemistry, biology and physics. Judging from his work approach, a strong collection of general texts, conceptual works, and reports from theory oriented meetings would be necessary. Since he relied on data from experimentalists to develop his models and theories, he would also require an index of potential data sources—people, labs, data sets, and archives. Accommodating the molecular networker would expand both collections and access capabilities of the system, but with this growth the precision retrieval mechanisms would need to be maintained to continue to adequately support Tom's practices. In other words, both content and functionality need to be founded on the way researchers work.

Distributed Knowledge

Researchers have independent and dependent ways of developing knowledge. Individuals rely on other people's knowledge to different degrees, and a researcher's level of dependent knowing varies from project to project. For some purposes, it may be adequate to have contact with an expert in an area and depend on that person for consultation. Another project may require recruiting a collaborator as a full member of a research team. The kind of dependent knowing practiced by participatory researchers works as an important strategy for combating overload, and it is also a technique that could be delegated to information specialists or, in some cases, to automated systems. The activities of interdisciplinary information intermediaries would concentrate on the periphery of a problem area, informing researchers of new work as well as relevant people, places, and activities outside the core of their problem area. As is currently the case with informal knowledge dependency, professional intermediaries could free researchers from some information probing and gathering, allowing them to concentrate on essential core maintenance work and the actual acquisition and application of new knowledge from the periphery. This is one example of skilled interdisciplinary articulation work that could assist individual projects and, over time, advance research fronts more generally.

Generalists tend to be more autonomous knowers. They are likely to read extensively or attend a seminar before consulting a colleague about specific details. Learning is a key component of their research process, therefore their information environments need to be conducive to self-education. In some ways, the information needs of generalists parallel those of students who are developing background in chosen field of study. For pedagogical purposes, derivative works, such as textbooks, handbooks, and review literature, are important counterparts to the masses of scholarly research reports. Research centers and libraries that wish to support interdisciplinary work need to recognize the essential role of advanced self-education when developing policies for collecting information resources. The types of materials that are studied and consulted by generalists are good candidates for working digital collections that can be shared by many and accessed remotely. Reading practices provide fundamental indictors of what should be brought together in digital libraries. In addition to the basic collections of multidisciplinary periodicals and general texts, content can be selected and centralized around the spectra of literature used by hybrid communities to create virtual problem centered resources. The different ways that researchers read also suggest that technologies can be implemented to exploit the various ways texts are used and managed during the course of research (Bishop 1999, Palmer and Neumann in press).

Another type of knowing that has not yet been discussed is "knowing what to forget." The Photosynthesis Specialist suggested that we must develop our "forgeteries" at the same time we build our memories.[7] In a similar sense, we can build filters and layers into our information systems to conform to a memory/forgetery model. As we study scientists to find out what is currently important to "know," we should also take cues from them about what can be preserved in the archival forgetery. As problem centered digital information environments are developed, current research models can be superimposed over older ones. For instance, a cross-disciplinary resource on computer vision could emphasize specialized material from artificial intelligence, graphics, and psychophysics in the foreground while disciplinary root material from computer science, engineering, and physiology is receded into the background. Through hypertextual layering, digital resources can retain the core and its ancestral categories of knowledge in the forgetery while allowing the new hybrids to be displayed on the surface. Simplifying access to information by emphasizing the most relevant bodies of material also helps resolve at least one aspect of the "computer-as-time-pit" problem.

There are important differences between the information seeking behavior of interdisciplinary researchers and those in more classical academic fields (Bates 1996), and there is evidence that more researchers take a horizontal or boundary crossing approach to their research than is presently acknowledged (Pierce 1999, Palmer 2001). If a large percentage of scholars are actively accumulating information across numerous domains, then our existing information systems and access tools are adequate for a much smaller population than imagined. Undoubtedly, many users will still have a need for disciplinary approaches to information. The digital information environment gives us the capability to create flexible systems that can foreground either the periphery or the core, whichever framework is best suited to the researcher's problem and approach. Since the academic reward system discourages scientists from taking on information compilation projects, information professionals need to initiate collaborative arrangements with domain experts to produce high quality problem-centered digital research tools.

Realizing Information Export

Interdisciplinary research requires information retrieval, current awareness, and new learning within and across meaningful domains of knowledge. Since the salience of domains is mutable, organizational structures must be flexible, and since exchange is the essence of interdisciplinary work, information needs to be easily accessible across domains. However, scientists are inclined to seek and gather information but not to actively share across research communities. Therefore, interdisciplinary information systems need to include mechanisms that generate export across boundaries. The platforms for exchange in systems and services should match the types of interactions that are important to researchers. Pahre's (1995) claim that research communities form around results, rather than data or concepts, seems plausible within the context of interdisciplinary science. Stores of data do little to aid current awareness or learning in a new field. The role of concepts is less clear. While researchers frequently mentioned borrowing concepts, the most telling cases of fruitful exchange came from access to results. Undeveloped results, what the Computational Neuroscientist called findings "with no deep ideas attached yet," are the most pliable. Exposure to raw results allows scientists to think about them in relation to their own research problem and methods before they are formulated into a paper that has been composed to fit the profile of a discipline-based journal or the preferences of an editorial board. This is why the researchers found specialized conferences so worthwhile; they funnel early results to those who can make the most use of them.

In light of these findings, it appears that the information professions have been prematurely focused on how to control the quality of digital information yet have not fully understood what constitutes quality. Research progress is dependent on many types of exchange, and information providers need to be meeting researchers' needs for diverse information, controlling for "quality" when feasible and necessary but also finding ways to manage other types of valuable information. Initiatives devoted to networking electronic data archives have been a progressive aspect of digital library development. Raw data and unprocessed results are increasingly available online and will become more valuable as they are linked to related results in preliminary research reports and refereed journal articles. As we upgrade information service organizations, it would be a mistake to continue to emphasize only the published product, or the electronic equivalent. We will need to develop new standards and criteria for the presentation of data and unprocessed results and create platforms for interchange and copresence around these valuable resources.

While it may be true that interactive communities do not organize around ideas and theories (Pahre 1995), the "pattern recognition" activities highlighted by a number of researchers suggest that conceptual relationships are important information entities. The patterns and topologies that researchers are able to discern within stores of information or across fields of information contribute to the process of discovery and analysis. Concepts are also used as communication tools for relating difficult ideas to multidisciplinary audiences. At present, concept-based information resources are rare. Culturally ingrained concepts are used as access points in bibliographic databases, and broad cross-disciplinary and conceptual classifications have begun to appear on Internet gateway indexes. Concept-based reference works and bibliographies are becoming more common, perhaps as a response to the growth of interdisciplinary knowledge. Mapping concepts across problem domains can help to identify, and potentially predict, broader knowledge structures that are not bound by specialization or the import-dominated exchange practices favored by scientific communities. In addition, tracing and analyzing the development and movement of concepts across domains is one way to study how pidgin and creole language cultures evolve, knowledge that could be used to further exploit information retrieval systems for discovery and synthesis. For example, the Vision Specialist spoke of how the term "accommodation" migrated from studies of the human eye in psychophysics to ocular machines in artificial intelligence. This concept could be traced much further to include the use of accommodation in linguistic theory, spatial orientations in architecture, and adaptation in biology. Likewise, it is possible that Tom might benefit from knowing how oscillation is applied to the notion of noise in information theory.

There is an interesting discord between the researchers' ambivalence about their audiences and their need for intellectual proximity to like-minded scientists. The emphasis in daily work routines, as well as in reward structures more generally, is clearly on importing information that can help produce new and valuable knowledge. At the same time, there is no doubt that the scientists benefit from exposure to information from the periphery, and it follows that active information dissemination of the right information to the right audiences could do much to advance new intellectual associations. Connecting conceptual terminology across information products through thesauri and other means can help the cause of cross-disciplinary dissemination, as can improved capabilities for searching multiple files and databases. Thus far, most of the existing online database systems that provide this function group files in ways that tend to reinforce traditional disciplinary delineations, and the very large multidisciplinary databases that are available offer limited assistance or capabilities for searching subjects across disciplines.

As information professionals begin to identify meaningful user groups based on problem areas and broader, overarching conceptual territories, they will become the experts on audience. Potential audiences can work as focal points for systematic information export services that engineer freer exchange across boundaries. Audience-based services could create linguistic and digital links between existing information sources and actively produce resources and tools that are configured around problems. The logical next step is to proactively disseminate those resources to all potential audiences. Information specialists may never become official authorities on authoritative information in either disciplines or hybrid domains (Wilson 1983), but they work from the best vantage point for monitoring the landscape of research activity, and therefore to build problem and audience centered systems that produce valuable information leeway for researchers.

Interdisciplinary research progresses through the accumulation of knowledge across boundaries. There are many ways of practicing interdisciplinary research, but they all depend on boundary crossing information work. Information fuels the generation of new interdisciplinary knowledge. Organizing information around research problems and opening up communication channels between problem-based research communities can create a robust information environment for interdisciplinary science. To capitalize on dynamic information environments, however, interdisciplinary researchers must work within structures that reinforce their multiple roles and diverse approaches. Dedicated support for the technical and the cultural aspects of accumulation work can enable interdisciplinarity in all kinds of research settings, activate the dialectic necessary for the synthesis of knowledge, and clear the path for future scientific discoveries.

---

[1] The literature on electronic communication in science and scholarship has proliferated since this study was conducted. For a sampling of the research, trends, and issues being debated see Harnad 1991, Crawford 1996, Hurd 1996, Ginsparg 1997, Walsh and Bayma 1997, and Bloom 2000.

[2] Turner (1994) uses the concept of the "intelligent mediator" to refer to machines developed for the coordination of data flow in electronic collaboratories. I use it here strictly in the human sense, in accordance with the idea that delegated work may be performed by machines or people (Latour 1992).

[3] See Palmer (1996) for a collection of papers addressing the library and interdisciplinary inquiry.

[4] Precision and recall are standard information retrieval effectiveness measures. Precision refers to the number of documents judged to be relevant out of the total number retrieved. Recall is the number of relevant documents retrieved out of the total number of relevant documents in the database or collection searched (Tonta 1992).

[5] The Annual Reviews series is an example of review publications within the social sciences, physical sciences, and the biological and medical sciences that offers searchable Web products. The publications are advertised as "the intelligent synthesis of the scientific literature."

[6] The user-centered revolution in library and information science is widely acknowledged. See for example Wilson 1981, Dervin and Nilan 1986, Taylor 1991, Savolainen 1993, Fidel 1994, and Nahl 1996.

[7] He attributed this idea to Peter Mitchell (1920-1992), the Nobel Prize winning chemist who established the validity of the chemiosmotic hypothesis.

APPENDIX

CO-WORD MAPS

As a relational indicator, the co-word technique traces connections between concepts, methods, and problems. The Leximappe co-word program creates visual representations of the associations between keywords, producing a cluster map that illustrates the relationships between research themes and a series of theme maps detailing the subnetworks within each theme.

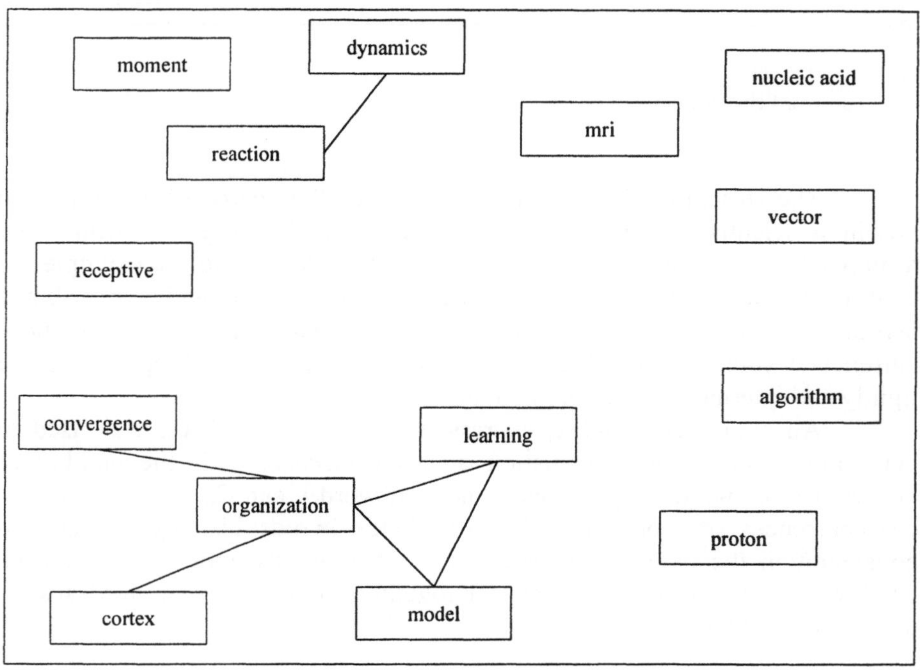

Co-word cluster map A

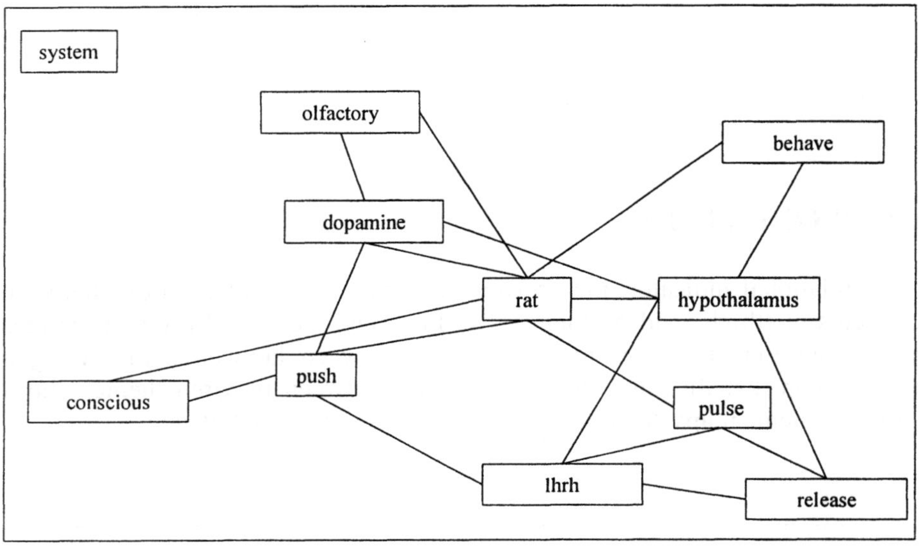

Co-word cluster map B

The structure of the cluster map reveals the degree of consolidation within a scientist's research domain; the words labeling the themes are arbitrary tags. The two sample cluster maps provided above are examples of contrasting cases. Map A has a broad range of themes and a distributed layout, indicating that this scientist's research domain has emerging or non-entrenched connections. Map B represents a scientist working in a more tightly and internally linked problem area.

An equivalence index, expressed as $E_{ij} = C_{ij}^2 / C_i{*}C_j$, was used to determine a normalized calculation of co-occurrence. C_i is the number of occurrences of word i, C_j the occurrences of word j, and C_{ij} is the number of co-occurrences of words i and j. The index avoids bias by taking into consideration the overall frequency of the two words under consideration. E_{ij}=1 when two words always appear together, or 0 when one word always excludes the other.

EXPORT PROFILES

The export profiles were developed to identify the scientists who had reached the broadest audience. The profiles were based on citations to a researcher's publications for the designated time period (1989-1993) as documented through searches of the ISI (Institute for Scientific Information) citation databases—SciSearch, Social SciSearch, and Arts & Humanities Search. Frequency counts were made of the citing journals for each researcher. As part of the process of developing the profiles, I compiled a directory of the 368 citing journals. Each entry in the directory contained the journal title, the Library of Congress subject headings, and other descriptive information from *Ulrich's International Periodicals Directory* and standard bibliographic records. Most entries also included a narrative description of the journal provided by the publisher and an account of the indexing and abstracting services that cover the title. Based on this composite data, a label indicating disciplinary integration was assigned to each journal: multidisciplinary, disciplinary/broad subject, subdisciplinary/fission, interdisciplinary/fusion, or problem-centered. Individual export profiles consisted of the following information:

number of times a specific journal cited the researcher
ratio of number of journal titles to total number of citations
subject areas associated with citing journals
ratio of number of subject areas to total number of citing journals
databases in which citations were found
breakdown of citations by assigned general subject designation

The two sample export profiles provided below are examples of contrasting cases. Profile A is an example of a researcher with a broad multidisciplinary audience, as displayed by citations in both SciSearch and SocialScisearch and high representation in interdisciplinary journals. Profile B is an example of a researcher with a narrow audience, shown by the concentration of citations in a limited number of disciplinary or subdisciplinary journals.

Analytical Biochemistry	1	biochemistry	d
Anatomical Record	1	anatomy	b
Annual Rev of Neuroscience	1	neuroscience	d
Artificial Intelligence Review	1	ai	d
Behavioral & Neural Biology	1	neurobiology	d
Beavioural Brain Research	1	neuropsychology	d
Biological Cybernetics	2	cybernetics, biological	d
Brain, Behavior & Evolution	1	neuropsychology	d
Brain Research	1	brain (multidisc)	d
Brain Research Bulletin	1	neuroscience	d
Comp App in Biosciences	1	life sciences—comp prog	c
Electrophoresis	1	electrophoresis	e
Human Movement Science	1	human mechanics	d
IEEE Trans Neural Networks	1	neural networks	e
International J Neuroscience	2	neuroscience	d
J Cognitive Neuroscience	1	neuropsychology	d
J Comparative Neurology	1	neurology	b
J Neuropsyiology	1	neurophysiology	d
J Neuroscience	2	neuroscience	d
J Theoretical Bio	1	biology	b
Microscopy Res & Technique	1	electron microscopy	e
Neural Computation	1	artificial intelligence	e
Neural Networks	2	neural networks	e
Proc of NAS	1	science	a
Proc Roy Soc London, Ser B	1	science	b
Revue Neurologique	1	neurology	b
Science	2	science	a
Trends in Neuroscience	3	neuroscience	d
Number of total items	35	Titles/Items	0.80
Number of titles	28		
Subject areas: science; anatomy; biology; neurology; life science—computer programs; biochemistry; neuroscience; neurobiology; neuropsychology; cybernetics; brain; neurophysiology; artificial intelligence; electrophoresis; neural networks; electron microscopy			
Number of subject areas	16	Subjects/Titles	0.57
Databases:		scisearch and social scisearch	
	#	%	
a-Multidisciplinary	3	9	
b-Disciplinary/broad subject	5	14	
c-Subdisciplinary/subdivision	1	3	
d-Interdisciplinary/fusion	20	57	
e-Problem-centered	6	17	

Export profile A

Functional Analysis and Its Applications	1	math—functional analysis	c
J Mathematical Physics	2	mathematical physics	c
Nuclear Physics B	1	nuclear physics	c
Physical Review A	1	nuclear physics	c
Physical Review D	1	nuclear physics	c
Physical Review B	1	nuclear physics	c
Progress of Theoretical Physics Supplement	1	physics	b
Number of total items	8	Titles/Items	0.88
Number of titles	7		
Subject areas: physics; math—functional analysis; mathematical physics; nuclear physics			
Number of subject areas	4	Subjects/Titles	0.57
Databases:		scisearch	
	#	%	
Multidisciplinary	0	0	
Disciplinary/broad subject	1	12.5	
Subdisciplinary/subdivision	7	87.5	
Interdisciplinary/fusion	0	0	
Problem-centered	0	0	

Export profile B

REFERENCES

Allen Thomas J. 1966. *Managing the Flow of Scientific and Technical Information.* Cambridge, MA: MIT Press.

Alpert, Daniel. 1985. "Performance and Paralysis: The Organizational Context of the American Research University." *Journal of Higher Education* 56 (3): 241-81.

American Psychological Association. 1963. *Reports of the American Psychological Association's Project on Scientific Information Exchange in Psychology.* Washington, D.C.: American Psychological Association.

Anbar, Michael. 1986. "The 'Bridge Scientist' and His Role." In *Interdisciplinary Analysis and Research: Theory and Practice of Problem-Focused Research and Development*, edited by Daryl E. Chubin, Alan L. Porter, and Frederick A. Rossini. Mt. Airy, MD: Lomond.

Bates, Marcia J. 1989. "The Design of Browsing and Berrypicking Techniques for the Online Search Interface." *Online Review* 13 (5): 407-24.

Bates, Marcia J. 1996. "Learning about the Information Seeking of Interdisciplinary Scholars and Students." *Library Trends* 45 (2): 155-64.

Bawden, David. 1986. "Information Systems and the Stimulation of Creativity." *Journal of Information Science* 12 (5): 203-16.

Beam, Robert D. 1983. "Fragmentation of Knowledge: An Obstacle to Its Full Utilization." In *The Optimum Utilization of Knowledge: Making Knowledge Serve Human Betterment*, edited by Kenneth E. Boulding and Lawrence Senesh. Boulder, CO: Westview.

Becher, Tony. 1989. *Academic Tribes and Territories: Intellectual Enquiry and the Cultures of Disciplines.* Milton Keynes, UK: Society for Research into Higher Education and Open University Press.

Becher, Tony. 1990. "The Counter-Culture of Specialization." *European Journal of Education* 25 (3): 333-46.

Becker, Betsy Jane. 1996. "Discourse Synthesis in Meta-Analysis." *Social Epistemology* 10 (1): 89-105.

Becker, Howard S. 1970. *Sociological Work: Method and Substance.* Chicago: Aldine.

Becker, Howard S. 1982. *Art Worlds.* Berkeley: University of California Press.

Becker, Howard S. 1998. *Tricks of the Trade: How to Think about Your Research While You're Doing It.* Chicago: University of Chicago Press.

Bernier, Charles L., and A. Neil Yerkey. 1979. "Message Diffusion." In *Cogent Communication: Overcoming Reading Overload.* Westport, CN: Greenwood Press.

Bernstein, Basil B. 1977. *Class, Codes, and Control.* Rev. ed. Vol. 3. London: Routledge and Kegan Paul.

Bernstein, Basil B. 1982. "Codes, Modalities, and the Process of Cultural Reproduction: A Model." In *Cultural and Economic Reproduction in Education*, edited by M. W. Apple. Boston: Routledge and Kegan Paul.

Bishop, Ann Peterson. 1999. "Document Structure and Digital Libraries: How Researchers Mobilize Information in Journal Articles." *Information Processing and Management* 35 (3): 255-79.

Bloom, Floyd E. 2000. "Lunch Selections Expanding." *Science*, 4 February, 801.

Borgman, Christine. L., ed. 1990. *Scholarly Communication and Bibliometrics.* Newbury Park, CA: Sage.

Boulding, Kenneth E. 1968. "Knowledge as a Commodity." In *Beyond Economics: Essays on Society, Religion, and Ethics.* Ann Arbor: University of Michigan Press.

Boulding, Kenneth E. 1983. "The Optimum Utilization of Knowledge: Some Central Concepts." In *The Optimum Utilization of Knowledge: Making Knowledge Serve Human Betterment*, edited by Kenneth E. Boulding and Lawrence Senesh. Boulder, CO: Westview.

Bourdieu, Pierre. 1975. "The Specificity of the Scientific Field and the Social Conditions for the Progress of Reason." *Social Science Information* 14 (6): 19-47.

Bowker, Geoffrey C. 1993. "How to Be Universal: Some Cybernetic Strategies, 1943-1970." *Social Studies of Science* 23 (1): 107-27.

Boyarin, Jonathan. 1993. *The Ethnography of Reading.* Berkeley: University of California Press.

Bradford, Samuel. C. 1953. *Documentation.* 2d ed. London: C. Lockwood.

Broudy, Harry S. 1970. "On Knowing With." In *Proceedings of the Philosophy of Education Society: Proceedings of the Twenty-sixth Annual Meeting of the Philosophy of Education Society—Atlanta, March 21-24 , 1970*, edited by Harold B. Dunkel. Edwardsville, IL: Studies in Philosophy and Education

Brown, John Seely, and Paul Duguid. 1991. "Organizational Learning and Communities-of-Practice: Toward a Unified View of Working, Learning, and Innovation." *Organization Science* 2 (1): 40-57.

Callon, Michael, Jean-Pierre Courtail, William A. Turner, and Serge Bauin. 1983. "From Translations to Problematic Networks: An Introduction to Co-word Analysis." *Social Science Information* 22 (2): 191-235.

Callon, Michael, John Law, and Arie Rip. 1986. *Mapping the Dynamics of Science and Technology: Sociology of Science in the Real World.* Basingstoke, Hampshire: Macmillan.

Campbell, Donald. 1969. "Ethnocentrism of Disciplines and the Fish-Scale Model of Omniscience." In *Interdisciplinary Relationships in the Social Sciences*, edited by Muzafer Sherif and Carolyn Sherif. Chicago: Aldine.

Carlson, W. Bernard. 1991. *Innovation as a Social Process: Elihu Thomson and the Rise of General Electric, 1870-1900.* Cambridge: Cambridge University Press.

Chang, Shan-Ju, and Ronald E. Rice. 1993. "Browsing: A Multidimensional Framework." *Annual Review of Information Science and Technology* 28: 231-76.

Chen, Hsinchun, Tak Yim, David Fye, and Bruce Schatz. 1995. "Automatic Thesaurus Generation for an Electronic Community System." *Journal of the American Society for Information Science* 46 (3): 175-93.

Chen, Hsinchun, Tobun D. Ng, Joanne Martinez, and Bruce Schatz. 1997. "A Concept Space Approach to Addressing the Vocabulary Problem in Scientific Information Retrieval: An Experiment on the Worm Community System." *Journal of the American Society for Information Science* 48 (1): 17-31.

Choi, Jin M. 1988. "Citation Analysis of Intra- and Interdisciplinary Communication Patterns of Anthropology in the U.S.A." *Behavioral and Social Sciences Librarian* 6 (3/4): 65-84.

Chubin, Daryl E. 1976. "The Conceptualization of Scientific Specialties." *Sociological Quarterly* 17 (4): 448-76.

Chubin, Daryl E. 1983. *Sociology of Sciences: An Annotated Bibliography on Invisible Colleges, 1972-1981.* New York: Garland.

Chubin, Daryl E., Alan L. Porter, and Frederick A. Rossini, eds. 1986. *Interdisciplinary Analysis and Research: Theory and Practice of Problem-Focused Research and Development.* Mt. Airy, MD: Lomond.

Chubin, Daryl E., Alan L. Porter, and Frederick A. Rossini. 1986. "Interdisciplinarity: How Do We Know Thee?—A Bibliographic Essay." In *Interdisciplinary Analysis and Research: Theory and Practice of Problem-Focused Research and Development*. Mt. Airy, MD: Lomond.

Clark, Burton R. 1995. *Places of Inquiry: Research and Advanced Education in Modern Universities.* Berkeley: University of California Press.

Clarke, Adele, and Joan H. Fujimura, eds. 1992. "What Tools? Which Jobs? Why Right?" In *The Right Tools for the Job: At Work in Twentieth-Century Life Sciences.* Princeton: Princeton University Press.

Clifford, James, and George E. Marcus. 1986. *Writing Culture: The Poetics and Politics of Ethnography.* Berkeley: University of California Press.

Cole, Charles. 1994. "Operationalizing the Notion of Information as a Subjective Construct." *Journal of the American Society for Information Science* 45 (7): 465-76.

Compton, Bertita E. 1966. "A Look at Conventions and What They Accomplish." *American Psychologist* 21 (2): 176-83.

Cooper, Harris M. 1989. *Integrating Research: A Guide for Literature Reviews*. 2d ed. Newbury Park, CA: Sage.

Corbin, Juliet M., and Anselm L. Strauss. 1993. "The Articulation of Work through Interaction." *Sociological Quarterly* 34 (1): 71-83.

Courtial, Jean-Pierre, and John Law. 1989. "A Co-Word Study of Artificial Intelligence." *Social Studies of Science* 19 (2): 301-11.

Coyne, Jerry A. 1995. "Dobzhansky Revisited." *Science*, 18 August, 991-92.

Crane, Diana. 1969. "Social Structure in a Group of Scientists: A Test of the 'Invisible College' Hypothesis." *Sociological Review* 34 (3): 335-52.

Crane, Diana. 1972. *Invisible Colleges*. Chicago: University of Chicago Press.

Crawford, Susan Y. 1971. "Informal Communication among Scientists in Sleep Research." *Journal of the American Society for Information Science* 22 (5): 301-10.

Crawford, Susan Y. 1996. "Scientific Communication and the Growth of Big Science." In *From Print to Electronic: The Transformation of Scientific Culture*. Medford, NJ: Information Today.

Cronin, Blaise. 1982. "Invisible Colleges and Information Transfer: A Review and Commentary with Particular Reference to the Social Sciences." *Journal of Documentation* 38 (3): 212-36.

Cronin, Blaise. 1995. *The Scholar's Courtesy: The Role of Acknowledgements in the Primary Communication Process*. London: Taylor Graham.

Davies, Roy. 1989. "The Creation of New Knowledge by Information Retrieval and Classification." *Journal of Documentation* 45 (4): 273-301.

de Mey, Marc. 1982. *The Cognitive Paradigm: Cognitive Science, a Newly Explored Approach to the Study of Cognition Applied in an Analysis of Science and Scientific Knowledge*. Dordrecht: D. Reidel.

Dervin, Brenda, and Michael Nilan. 1986. "Information Needs and Uses." *Annual Review of Information Science and Technology* 21: 3-33.

Dogan, Mattei, and Robert Pahre. 1990. *Creative Marginality: Innovation at the Intersections of Social Sciences*. Boulder, CO: Westview Press.

Dooris, Michael J., and James S. Fairweather. 1994. "Structure and Culture in Faculty Work: Implications for Technology Transfer." *Review of Higher Education* 17 (2): 161-77.

Ellis, David. 1993. "Modeling the Information-Seeking Patterns of Academic Researchers: A Grounded Theory Approach." *Library Quarterly* 63 (4): 469-86.

Fidel, Raya. 1994. "User-Centered Indexing." *Journal of the American Society for Information Science* 45 (8). 572-76.

Fisher, Donald. 1990. "Boundary Work and Science: The Relation between Power and Knowledge." In *Theories of Science in Society*, edited by Susan E. Cozzens and Thomas F. Gieryn. Bloomington: Indiana University Press.

Fujimura, Joan H. 1987. "Constructing 'Do-able' Problems in Cancer Research: Articulating Alignment." *Social Studies of Science* 17 (2): 257-93.

Fuller, Steve. 1988. "Disciplinary Boundaries: A Conceptual Map for the Field." In *Social Epistemology*. Bloomington: Indiana University Press.

Fuller, Steve. 1993. "The Position: Interdisciplinarity." In *Philosophy, Rhetoric, and the End of Knowledge*. Madison: University of Wisconsin Press.

Galison, Peter. 1996. "Computer Simulations and the Trading Zone." In *The Disunity of Science: Boundaries, Contexts, and Power*, edited by Peter Galison and David J. Stump. Stanford, CA: Stanford University Press.

Garvey, William D. 1979. *Communication: The Essence of Science*. Oxford: Pergamon Press.

Garvey, William D., and Belver C. Griffith. 1964. "The Effect of Convention Presentations on Information Exchange Behavior and Subsequent Research." *Proceedings of the American Documentation Institute* 1: 201-13.

Garvey, William D., and Belver C. Griffith. 1968. "Informal Channels of Communication in the Behavioral Sciences: Their Relevance in the Structuring of Formal or Bibliographic Communication." In *The Foundations of Access to Knowledge*, edited by Edward B. Montgomery. Syracuse, NY: Syracuse University.

Geertz, Clifford. 1983. *Local Knowledge: Further Essays in Interpretive Anthropology*. New York: Basic Books.

Gerson, Elihu M. 1983. "Scientific Work and Social Worlds." *Knowledge: Creation, Diffusion, Utilization* 4 (3): 357-77.

Gibbons, Michael, Camille Limoges, Helga Nowotny, Simon Schwartzman, Peter Scott, and Martin Trow. 1994. *The New Production of Knowledge: The Dynamics of Science and Research in Contemporary Societies*. London: Sage Publications.

Gieryn, Thomas F. 1983. "Boundary-Work and the Demarcation of Science from Non-Science: Strains and Interests in Professional Ideologies of Scientists." *American Sociological Review* 48 (6): 781-95.

Gieryn, Thomas F. 1995. "Boundaries of Science." In *Handbook of Science and Technology Studies* , edited by Sheila Jasanoff, Gerald E. Markle, James C. Petersen, and Trevor Pinch. Thousand Oaks, CA: Sage.

Gieryn, Thomas F. 1999. *Cultural Boundaries of Science: Credibility on the Line*. Chicago: University of Chicago Press.

Ginsparg, Paul. 1997. "The Impact of Electronic Publishing on the Academic Community." In *The Impact of Electronic Publishing on the Academic Community: An International Workshop Organized by the Academia Europaea and the Wenner-Gren Foundation*, edited by Ian Butterworth. London: Portland. Available at: http://tiepac.portlandpress.co.uk/books/online/tidpac/session1/ch7.html (visited 7.7.01)

Glaser, Barney G., and Anselm Strauss. 1967. *The Discovery of Grounded Theory*. Chicago: Aldine.

Granovetter, Mark S. 1973. "The Strength of Loose Ties." *American Journal of Sociology* 78 (6): 1360-80.

Griffith, Belver C., Marilyn J. Jahn, and A. James Miller. 1971. "Informal Contacts in Science: A Probabilistic Model for Communication Processes." *Science*, 9 July, 164-66.

Griffith, Belver. C., and A. James Miller. 1970. "Networks of Informal Communication among Scientifically Productive Scientists." In *Communication among Scientists and Engineers*, edited by Carnot E. Nelson and Donald K. Pollock. Lexington, MA: D. C. Heath.

Haas, Stephanie W. 1997. "Disciplinary Variation in Automatic Sublanguage Term Identification." *Journal of the American Society for Information Science* 48 (1): 67-79.

Harnard, Stevan. 1990. "Scholarly Skywriting and the Prepublication Continuum of Scientific Inquiry." *Psychological Science* 1 (6): 342-43.

Harper, Douglas. 1992. "Small N's and Community Case Studies." In *What Is a Case: Exploring the Foundations of Social Inquiry*, edited by Charles C. Ragin and Howard S. Becker. Cambridge: Cambridge University Press.

Harwood, Jonathan. 1993. *Styles of Scientific Thought: The German Genetics Community, 1900-1933*. Chicago: University of Chicago Press.

Hayles, N. Katherine. 1991. "Introduction: Complex Dynamics in Literature and Science." In *Chaos and Order: Complex Dynamics in Literature and Science*. Chicago: University of Chicago Press.

Hiltz, Starr Roxanne. 1984. *Online Communities: A Case Study of the Office of the Future*. Norwood, NJ: Ablex.

Hoch, Paul K. 1987. "Institutional versus Intellectual Migration in the Nucleation of New Scientific Specialties." *Studies in History and Philosophy of Science* 18 (4): 481-500.

Hudson, Kenneth. 1977. *The Dictionary of Diseased English*. New York: Harper and Row.

Hurd, Julie M. 1992. "Interdisciplinary Research in the Sciences: Implications for Library Organization." *College and Research Libraries* 53 (4): 283-97.

Hurd, Julie M. 1996. "Models of Scientific Communication Systems." In *From Print to Electronic: The Transformation of Scientific Culture*. Medford, NJ: Information Today.

Jantsch, Erich. 1980. "Interdisciplinarity: Dreams and Reality." *Prospects* 10 (3): 304-12.

Klein, Julie Thompson. 1990. *Interdisciplinarity: History, Theory, and Practice*. Detroit: Wayne State University.

Klein, Julie Thompson. 1993. "Blurring, Cracking, and Crossing: Permeation and the Fracturing of Discipline." In *Knowledges: Historical and Critical Studies in Disciplinarity*, edited by Ellen Messer-Davidow, David R. Shumway, and David J. Sylvan. Charlottesville: University Press of Virginia.

Klein, Julie Thompson. 1996. *Crossing Boundaries: Knowledge, Disciplinarities, and Interdisciplinarities*. Charlottesville: University Press of Virginia.

Klein, Julie Thompson. 2000. "The Discourse of Transdisciplinarity: An Expanding Global Field." In *Transdisciplinarity: Joint Problem Solving among Science, Technology and Society: An Effective Way for Managing Complexity*, edited by Julie Thompson Klein, Walter Grossenbacher-Mansuy, Rudolf Häberli, Alain Bill, Roland W. Schloz, and Myrtha Welti. Basel: Birkhauser.

Klein, Julie Thompson. Forthcoming. "History of Transdisciplinary Research: Contexts of Definition, Theory, and the New Discourse of Problem Solving." In *Encyclopedia of Life Support Systems*. Oxford: EOLSS.

Knorr-Cetina, Karin. 1981. *The Manufacture of Knowledge*. Oxford: Pergamon Press.

Kuhn, Thomas. 1962. *The Structure of Scientific Revolutions*. Chicago: University of Chicago Press.

Latour, Bruno. 1987. *Science in Action: How to Follow Scientists and Engineers through Society*. Cambridge, MA: Harvard University Press.

Latour, Bruno. 1992. "Where Are the Missing Masses? The Sociology of a Few Mundane Artifacts." In *Shaping Technology/Building Society: Studies in Sociotechnical Change*, edited by Wiebe E. Bijker and John Law. Cambridge: MIT Press.

Latour, Bruno, and Steve Woolgar. 1979. *Laboratory Life: The Social Construction of Scientific Facts*. London: Sage.

Lave, Jean, and Etienne Wenger. 1991. *Situated Learning: Legitimate Peripheral Participation*. Cambridge: Cambridge University Press.

Law, John. 1994. *Organizing Modernity*. Oxford: Blackwell Publishers.

Lemert, Charles C. 1990. "Depth as a Metaphor for the Major: A Postmodernist Challenge." Association of American Colleges meeting. San Francisco, 11 January. Quoted in Julie Thompson Klein, *Crossing Boundaries: Knowledge, Disciplinarities, and Interdisciplinarities*. Charlottesville: University Press of Virginia, 1996.

Lenoir, Timothy. 1993. "The Discipline of Nature and the Nature of Disciplines." In *Knowledges: Historical and Critical Studies in Disciplinarity*, edited by Ellen Messer-Davidow, David R. Shumway, and David J. Sylvan. Charlottesville: University Press of Virginia.

Lin, Nan, William D. Garvey, and Carnot E. Nelson. 1970. "A Study of the Communication Structure of Science." Johns Hopkins University. Center for Research in Scientific Communication, report no. 10. Baltimore: Johns Hopkins University.

Long, F. A. 1971. "Interdisciplinary Problem-Oriented Research in the University." *Science*, 12 March, 961.

Löwy, Ilana. 1992. "The Strength of Loose Concepts: Boundary Concepts, Federative Experimental Strategies, and Disciplinary Growth." *History of Science* 30 (4): 371-96.

Mann, Thomas. 1993. *Library Research Models: A Guide to Classification, Cataloging, and Computers*. New York: Oxford University Press.

McCain, Katherine. 1986. "Cross-Disciplinary Citation Patterns in the History of Technology." In *Proceedings of the 49th ASIS Annual Meeting*, Chicago, Illinois, September 28-October 2, 1986. Edited by J. M. Hurd and C. H. Davis. Medford, NJ: Learned Information.

McClure, Charles R., Ann P. Bishop, Philip Doty, and Howard Rosenbaum. 1991. *The National Research and Education Network (NREN): Research and Policy Perspectives*. Norwood, NJ: Ablex.

McInnis, Raymond G. 1996. "Introduction: Defining Discourse Synthesis." *Social Epistemology* 10 (1): 1-25.

Merton, Robert K. 1973. *The Sociology of Science*. Chicago: University of Chicago Press.

Messer-Davidow, Ellen, David R. Shumway, and David J. Sylvan. 1993. "Disciplinary Ways of Knowing." In *Knowledges: Historical and Critical Studies in Disciplinarity*. Charlottesville: University Press of Virginia.

Mulkay, Michael. 1974. "Conceptual Displacement and Migration in Science." *Science Studies* 4 (3): 205-34.

Myers, Greg. 1993. "The Social Construction of Two Biologists' Articles." In *Knowledges: Historical and Critical Studies in Disciplinarity*, edited by Ellen Messer-Davidow, David R. Shumway, and David J. Sylvan. Charlottesville: University Press of Virginia.

Nahl, Diane. 1996. "The User Centered Revolution: 1970-1995." In *Encyclopedia of Microcomputers*, vol. 19. New York: Marcel Dekker. Available at: http://www2.hawaii.edu/~nahl/articles/user/user1toend_toc.html (visited 7.7.01)

National Research Council. (U.S.). Physics Survey Committee. 1986. *Scientific Interfaces and Technological Applications*. Physics through the 1990s. Washington, D.C.: National Academy Press.

Neill, S. D. 1989. "The Information Analyst as a Quality Filter in the Scientific Communication Process." *Journal of Information Science* 15 (1): 2-12.

Neisser, Ulric. 1976. *Cognition and Reality: Principles and Implications of Cognitive Psychology*. San Francisco: W. H. Freeman.

O'Reilly, Charles A. 1982. "Variations in Decision Makers' Use of Information Sources: The Impact of Quality and Accessibility of Information." *Academy of Management Journal* 25 (4): 756-71.

Oseman, Robert. 1989. *Conferences and Their Literature: A Question of Value*. London: Library Association.

Oxman, Andrew D. and Gordon H. Guyatt. 1993. "The Science of Reviewing Research." *Annals of the New York Academy of Sciences* 703: 125-34.

Pahre, Robert. 1995. "Positivist Discourse and Social Scientific Communities: Toward an Epistemological Sociology of Science." *Social Epistemology* 9 (3): 233-55.

Pahre, Robert. 1996. "Patterns of Knowledge Communities in the Social Sciences." *Library Trends* 45 (2): 204-25.

Paisley, William J., and Edwin B. Parker. 1967. *Scientific Information Exchange at an Interdisciplinary Behavioral Science Convention.* Stanford, CA: Stanford University, Institute for Communication Research.

Palmer, Carole L., ed. 1996. *Navigating among the Disciplines: The Library and Interdisciplinary Inquiry. Library Trends* 45 (2).

Palmer, Carole L., 2001. "The Information Connection in Scholarly Synthesis." In *Discourse Synthesis: Studies in Historical and Contemporary Social Epistemology*, edited by Raymond G. McInnis. Westport, CT: Praeger.

Palmer, Carole L., and Laura J. Neumann. In press. "The Information Work of Interdisciplinary Humanities Scholars: Exploration and Translation." *Library Quarterly* 72.

Palmer, Judith. 1991. "Scientists and Information: Using Cluster Analysis to Identify Information Style." *Journal of Documentation* 47 (2): 105-29.

Perkins, James A. 1973. *The University as an Organization*. New York: McGraw Hill.

Petrie, Hugh G. 1986. "Do You See What I See? The Epistemology of Interdisciplinary Inquiry." In *Interdisciplinary Analysis and Research: Theory and Practice of Problem-Focused Research and Development*, edited by Daryl E. Chubin, Alan L. Porter, and Frederick A. Rossini. Mt. Airy, MD: Lomond.

Pickering, Andrew, ed. 1992. *Science as Practice and Culture*. Chicago: University of Chicago Press.

Pierce, Sydney J. 1999. "Boundary Crossing in Research Literatures as a Means of Interdisciplinary Information Transfer." *Journal of the American Society for Information Science* 50 (3): 271-79.

Pinch, Trevor. 1990. "The Culture of Scientists and Disciplinary Rhetoric." *European Journal of Education* 25 (3): 295-304.

Polanyi, Michael. 1958. *Personal Knowledge: Towards a Post-Critical Philosophy*. London: Routledge and Kegan Paul.

Price, Derek J. de Solla. 1961. *Science since Babylon.* New Haven, CT: Yale University Press.

Ravetz, Jerome R. 1971. *Scientific Knowledge and Its Social Problems.* New York: Oxford University Press.

Rosenberg, Victor. 1967. "Factors Affecting the Preferences of Industrial Personnel for Information Gathering Methods." *Information Storage and Retrieval* 3 (3): 119-27.

Rosenthal, MaryLu C. 1994. "The Fugitive Literature." In *The Handbook of Research Synthesis,* edited by Harris Cooper and Larry V. Hedges. New York: Russell Sage Foundation.

Rosenthal, Robert. 1995. "Writing Meta-Analytic Reviews." *Psychological Bulletin* 118 (2): 183-92.

Sandstrom, Pamela. E. 1994. "An Optimal Foraging Approach to Information Seeking and Use." *Library Quarterly* 64 (4): 414-49.

Savolainen, Reijo. 1998. "Use Studies of Electronic Networks: A Review of Empirical Research Approaches and Challenges for Their Development." *Journal of Documentation* 54 (3): 332-51.

Seidman, I. E. 1991. *Interviewing As Qualitative Research: A Guide for Researchers in Education and the Social Sciences.* New York: Teachers College Press.

Shephard, F. C. 1983. "The Various Roles of Secondary Publications (Some Thoughts)." *International Journal of Micrographics and Video Technology* 2 (2): 101-4.

Skelton, Barbara. 1973. "Scientists and Social Scientists as Information Users: A Comparison of Results of Science User Studies with the Investigation into Information Requirements of the Social Sciences." *Journal of Librarianship* 5 (2): 138-56.

Small, Henry G. 1977. "A Co-citation Model of a Scientific Specialty: A Longitudinal Study of Collagen Research." *Social Studies of Science* 7 (2): 139-66.

Small, Henry G. 1978. "Cited Documents as Concept Symbols." *Social Studies of Science* 8 (3): 327-40.

Small, Henry. G., and Edwin Greenlee. 1980. "Citation Context Analysis and the Structure of Paradigms." *Journal of Documentation* 36 (3): 183-96.

Smith, Jack T. Jr. 1996. "Meta-Analysis: The Librarian as a Member of an Interdisciplinary Research Team." *Library Trends* 45 (2): 265-79.

Sproull, Lee, and Sara Kiesler. 1991. *New Ways of Working in the Networked Organization.* Cambridge, MA: MIT Press.

Star, Susan Leigh. 1989. *Regions of the Mind: Brain Research and the Quest for Scientific Certainty.* Stanford, CA: Stanford University Press.

Star, Susan Leigh, and James R. Griesemer. 1989. "Institutional Ecology, 'Translations' and Boundary Objects: Amateurs and Professionals in Berkeley's Museum of Vertebrate Zoology, 1907-39." *Social Studies of Science* 19 (3): 387-420.

Star, Leigh, and Anselm Strauss. 1998. "Layers of Silence, Arenas of Voice: The Dialogues between Visible and Invisible Work." In *Social Science, Technical Systems, and Cooperative Work: Beyond the Great Divide*, edited by Geoffrey Bowker, Susan Leigh Star, and William A. Turner. Mahwah, NJ: Lawrence Erlbaum.

Stites, Janet. 1993. "Bordercrossings: A Conversation in Cyberspace." *Omni* 16 (2): 39-48, 105.

Strauss, Anselm. 1978. "A Social World Perspective." *Studies in Symbolic Interaction* 1: 119-28.

Strauss, Anselm. 1988. "The Articulation of Project Work: An Organizational Process." *Sociological Quarterly* 29 (2): 163-78.

Strauss, Anselm, and Juliet Corbin. 1990. *The Basics of Qualitative Research: Grounded Theory Procedures and Techniques.* Newbury Park, CA: Sage.

Swanson, Don. R. 1986a. "Fish Oil, Raynaud's Syndrome, and Undiscovered Public Knowledge." *Perspectives in Biology and Medicine* 30 (1): 7-18.

Swanson, Don. R. 1986b. "Undiscovered Public Knowledge." *Library Quarterly* 56 (2): 103-18.

Swanson, Don. R. 1987. "Two Medical Literatures That Are Logically but Not Bibliographically Connected." *Journal of the American Society for Information Science* 38 (4): 228-33.

Swanson, Don. R. 1988. "Migraine and Magnesium: Eleven Neglected Connections." *Perspectives in Biology and Medicine* 31 (4): 526-57.

Taylor, Robert S. 1991. "Information Use Environments." *Progress in Communication Sciences* 10: 217-55

Tonta, Yasar. 1992. "Analysis of Search Failures in Document Retrieval Systems: A Review." *The Public-Access Computer Systems Review* 3 (1): 4-53.

Tracz, George. 1980. "Computerized Conferencing: An Eye-Opening Experience with EIES." *Canadian Journal of Information Science* 5 (May): 11-20.

Traweek, Sharon. 1988. *Beamtimes and Lifetimes: The World of High Energy Physicists.* Cambridge, MA: Harvard University Press.

Turner, Ralph H. 1991. "The Many Faces of American Sociology: A Discipline in Search of Identity." In *Divided Knowledges: Across Disciplines, Across Cultures*, edited by David Easton and Corinne Schelling. Newbury Park, CA: Sage.

Turner, William A. 1994. "What's in an R: InfoRmetrics or Infometrics?" *Scientometrics* 30 (2/3): 471-80.

Turoff, Murray, and Starr Roxanne Hiltz. 1980. "Electronic Information Exchange and Its Impact on Libraries." In *The Role of the Library in an Electronic Society*, edited by F. W. Lancaster. Proceedings of the 1979 Clinic on Library Applications of Data Processing. Urbana-Champaign: University of Illinois.

Ulrich's International Periodicals Directory, 1994-95. 33rd ed. 5 vols. New Providence, NJ: R.R. Bowker.

von Hippel, Eric. 1988. *The Sources of Innovation*. New York: Oxford University Press.

Walsh, John P., and Todd Bayma. 1997. "Computer Networks and Scientific Work." In *Culture of the Internet*, edited by Sara Kiesler. Mahwah, NJ: Lawrence Erlbaum.

Weick, Karl E. 1970. "The Twigging of Overload." In *People and Information,* edited by H. B. Pepinsky. New York: Pergamon Press.

Weingart, Peter, and Nico Stehr, eds. 2000. *Practising Interdisciplinarity*. Toronto: University of Toronto Press.

White, Howard D. 1994. "Scientific Communication and Literature Retrieval." In *Handbook of Research Synthesis*, edited by Harris Cooper and Larry V. Hedges. New York: Russell Sage Foundation.

White, Howard D. 1996. "Literature Retrieval for Interdisciplinary Syntheses." *Library Trends* 45 (2): 239-64.

White, James Boyd. 1987. "Intellectual Integration." *Issues in Integrative Studies* 5: 1-18.

Wilson, Patrick. 1983. *Second-Hand Knowledge: An Inquiry Into Cognitive Authority*. Westport, CT: Greenwood Press.

Wilson, Patrick. 1996. "Interdisciplinary Research and Information Overload." *Library Trends* 45 (2): 192-203.

Wilson, Thomas D. 1981. "On User Studies and Information Needs." *Journal of Documentation* 37 (1): 3-15.

NAME INDEX

Allen, Thomas J., 59 n1
Alpert, Daniel, 80-81, 131
American Psychological Association, 36
Anbar, Michael, 78-79

Bates, Marcia J., 32, 142
Bawden, David, 136-137
Bayma, Todd, 145 n1
Beam, Robert D., 8
Becher, Tony, 2, 13, 50, 70, 77
Becker, Betsy Jane, 137
Becker, Howard S., 14, 17, 27 n6
Bernier, Charles L., 59 n4
Bernstein, Basil B., 27 n4
Bishop, Ann Peterson, 142
Bloom, Floyd E., 145 n1
Borgman, Christine L., 14
Boulding, Kenneth E., vii, 135
Bourdieu, Pierre, 44
Bowker, Geoffrey C., 76
Boyarin, Jonathan, 59 n10
Bradford, Samuel C., 8
Broudy, Harry S., 86
Brown, John Seely, 112, 113, 129, 139

Callon, Michael, 14-15, 16, 25, 37, 47, 100
Campbell, Donald, 7, 62
Carlson, W. Bernard, 27 n3
Chang, Shan-Ju, 40
Chen, Hsinchun, 134
Choi, Jin M., 27 n7
Chubin, Daryl E., 7, 8, 13, 27 n7, 110
Clark, Burton R., 2
Clarke, Adele, 27 n6, 65
Clifford, James, 59 n10
Cole, Charles, 135
Compton, Bertita E., 36
Cooper, Harris M., 44
Corbin, Juliet M., 18, 99
Courtail, Jean-Pierre, 16
Coyne, Jerry A., 27 n5
Crane, Diana 8, 37, 125
Crawford, Susan Y., 125, 145 n1
Cronin, Blaise, 14, 16, 37, 125

Davies, Roy, 136, 137
de Mey, Marc, 135
Dervin, Brenda, 145 n6
Dogan, Mattei, 8, 64
Dooris, Michael J., 69-70
Duguid, Paul, 112, 113, 129, 139

Ellis, David, 17, 31

Fairweather, James S., 69-70
Fidel, Raya, 145 n6
Fisher, Donald, 1, 7-8, 10-11, 25, 27 n4
Fujimura, Joan H., 27 n6, 65, 98-100
Fuller, Steve, 1, 13, 95 n1, 97

Galison, Peter, 75-76
Garvey, William D., 32, 33, 37, 59 n6,137
Geertz, Clifford, 2, 19
Gerson, Elihu M., 11, 65
Gibbons, Michael, 1
Gieryn, Thomas F., 10-11
Ginsparg, Paul, 145 n1
Glaser, Barney G., 14, 17, 18
Granovetter, Mark S., 16
Greenlee, Edwin, 27 n9
Griesemer, James R., 11
Griffith, Belver C., 32, 33, 37, 59 n6, 125
Guyatt, Gordon H., 137

Haas, Stephanie W., 134
Harnad, Stevan, 145 n1
Harper, Douglas, 17
Harwood, Jonathan, 20, 76
Hayles, N. Katherine, 118, 121 n3
Hiltz, Starr Roxanne, 41, 57
Hoch, Paul K., 7
Hudson, Kenneth, ix
Hurd, Julie M., 27 n7, 145 n1

Jahn, Marilyn J., 125
Jantsch, Erich, 1

Kiesler, Sara, 59 n5
Klein, Julie Thompson, ix-x, 1-2, 7, 9, 10-11, 16-17, 20, 27 n1, 56, 67, 85-86, 95 n2, 133
Knorr-Cetina, Karin, 13
Kuhn, Thomas, 20, 81

Latour, Bruno, 9-10, 13, 14-15, 37, 53, 64, 100, 130, 145 n2
Lave, Jean, 49, 71, 129
Law, John, 14-15, 16, 20, 25, 37, 47, 100
Lemert, Charles C., 95 n2
Lenoir, Timothy, 25-26, 64-65, 113
Lin, Nan, 37
Long, F. A., 20
Löwy, Ilana, 75

Mann, Thomas, 135
Marcus, George E., 59 n10
McCain, Katherine, 27 n7
McClure, Charles R., 41, 59 n8
McInnis, Raymond G., 137, 139
Merton, Robert K., 70
Messer-Davidov, Ellen, 14
Miller, A. James, 33, 125
Mulkay, Michael, 7
Myers, Greg, 59 n11, 81-82

Nahl, Diane, 145 n6
National Research Council, 9
Neill, S. D., 59 n4
Neisser, Ulric, 135
Nelson, Carnot E., 37
Neumann, Laura J., 142
Nilan, Michael, 145 n6

O'Reilly, Charles A., 59 n2
Oseman, Robert, 59 n7
Oxman, Andrew D., 137

Pahre, Robert, 8, 64, 135, 143
Paisley, William J., 36
Palmer, Carole L., 142, 145 n3
Palmer, Judith, 27 n12
Parker, Edwin B., 36
Perkins, James A., 131
Petrie, Hugh G., 86-87
Pickering, Andrew, 139
Pierce, Sydney J., 142
Pinch, Trevor, 2, 70
Polanyi, Michael, 86

Porter, Alan L., 27 n7
Price, Derek J. de Solla, viii, 37, 125

Ravetz, Jerome R., 13, 49
Rice, Ronald E., 40
Rip, Arie, 14-15, 16, 25, 37, 47, 100
Rosenberg, Victor, 59 n1
Rosenthal, MaryLu C., 137
Rosenthal, Robert, 137
Rossini, Frederick A., 27 n7

Sandstrom, Pamela E., 15, 136, 137
Savolainen, Reijo, 145 n6
Seidman, I. E., 17
Shephard, F. C., 137
Shumway, David R., 14
Skelton, Barbara, 32, 41
Small, Henry G., 15, 27 n9
Smith, Jack T. Jr., 137
Sproull, Lee, 59 n5
Star, Susan Leigh, 11, 13, 139
Stehr, Nico, ix
Stites, Janet, 121 n3
Strauss, Anselm, 13-14, 17, 18, 65, 99, 100
Swanson, Don R., 136
Sylvan, David J., 14

Taylor, Robert S., 145 n6
Tonta, Yasar, 145 n4
Tracz, George, 57
Traweek, Sharon, 106-107
Turner, Ralph H., 1
Turner, William A., 15, 16, 145 n2
Turoff, Murray, 57

von Hippel, Eric, 59 n9, 129

Walsh, John P., 145 n1
Weick, Karl E., 93-94
Weingart, Peter, ix
Wenger, Etienne, 49, 71, 129
White, Howard D., 43, 136, 139
White, James Boyd, 69-70, 72, 85-86
Wilson, Patrick, 129, 145
Wilson, Thomas D. , 145 n6
Woolgar, Steve, 13

Yerkey, A. Neil, 59 n4

SUBJECT INDEX

acculturation, 129-130
accumulation work, 9-10, 25, 123
  and information, 133-134
  and literature, 137-138
articulation work, 99-101
audience(s), 50-56, 90
  general, 52-54
  multiple, 51
  see also literature

bibliometrics, 14-15
  and system design, 139
boundary crossing
  characteristics, 21
  and leeway, 99
  literature, 14-17
  work, 10-11
boundary crossing modes, 20-25
  and distributed knowledge, 141-142
  and interaction, 126
  and knowledge units, 91-92
  and leeway, 127-128
  see also collaborators, generalists, team leaders
boundary objects, 11
boundary trajection, 11, 124
boundary work, 10-11
  see also cross-disciplinary work
breadth vs. focus, 109-111
browsing, 40-41, 140

career paths, 77, 82, 106
centralization, 111-115
citation analysis, 15-17, 38-39, 119
  study methodology, 29
cognitive approaches, 76-79
collaboration, 102, 109-110, 115-118, 123-124
  additive model of, 116
  barriers to, 117-118
  and communication, 118
  expectations, 123-124
collaborative generalists, 107-108, 126-127
  and interaction, 127
  and leeway, 127-128
collaborators, 21-22
  and conferences, 36
  and generalists, 107-110
  information seeking of, 31
communication, 124-126
  and collaboration, 118
  and creolization, 75-76
  cross-cultural, 72-77
  cross-disciplinary, 56-57, 132-133
  cultural barriers to, 69-70
  difficulties of, 72-76
  electronic, 38, 57-58, 125-126
  and import, 32
  informal, 112-113
  and interdisciplinary institutions, 101
  and knowledge units, 8

and social worlds, 100
and translation, 130
vocabulary, 73-75
see also intermediaries, personal networks
communities, intellectual, 69-72
communities of practice, 49, 54, 58, 71
competition, 120
comprehension, 75, 85-87, 133
computer science, 63, 110
concepts, 134, 143-144
conferences, 35-37, 55
copresence, 129-131
core and scatter, 8-9, 15-16
and research focus/breath, 110-111
and scientific growth curve, 79-82
co-word analysis, 15-16, 38-39
creativity, 76, 78, 136-137
cross-disciplinary work
data and data sources in, 12
group interaction in, 26-27
and interdisciplinary institutions, 64-65
and language, 12
and methods, 12
and people, 11-12, 17-20
see also accumulation work
cross-disciplinary work, study of, 13-18
cultural and organizational context, 13-14
data collection in, 14-15, 17
and interdisciplinary research, 14
interviews for, 17-18

databases, 38, 40-41
delegation, 130
diffusion, 15
digital information
see electronic information
digital libraries, 142-143
disciplinary culture, 69-72
disciplines, ix, 2
boundaries of, 1
and boundary objects, 11
building of, 64-65, 113
categories of, 19, 26
and copresence, 129
core maintenance of, 82-83
and cross-disciplinary communication, 69-70
and cross-disciplinary training, 83-84
current research in, 62-63, 66
historical role of, 62
hybridization of, 8-9, 64
obsolescence of, 61-64
research in, 61-62, 114-115, 131
discovery, 135-138
distributed knowledge, 141-142

electronic information, 38, 40-41, 57-58, 105, 125, 135, 143
e-mail, 34, 57
equilibriums, 105-111
exchange of information, 30, 56-57
experimental research, 51-52, 106-107
export of information
and acculturation, 119
and import of information, 30
practices, 30fig, 50-56
processes, 30fig

focus vs. breadth, 109-111

generalists, 21 fig, 23-24, 76-77
and collaborative practices, 107-110
graduate students

see students, graduate
group work, 129-130

hybridization, 8-9, 64, 111-115

import of information
and export activities, 30
practices, 30fig, 32-50
processes, 30fig, 31-32
see also conferences, intermediaries, learning, literature, teaching
import/export balance, 119-120
information dissemination
see information export
information exchange, 30, 33-34, 56-57, 142-145
information export, 15-18, 119, 142-145
see also export activities
information gathering, 31-32
information import, 14-18, 31-32
see also import activities
information integration, 2
see also synthesis
information organization, 138-142
information overload, 93-95
information practices, 29-30
information processes, 29-30
see also import activities, export activities
information professionals
see information specialists
information quality, 143
information science
and language, 134-135
information seeking, 31-32
information services, 2, 29
for accumulation, 133
and discipline structure, 140
for discovery, 136
and distributed knowledge, 141-142
for information export, 143-144
and information sources, 32
and language, 135
information specialists, 133-135, 137-138, 141-145
information technology, 2, 17, 104-105, 125
and bibliometrics, 139
and cross-disciplinary work, 12
design, 138-141
and discovery, 136-137
and distributed knowledge, 140-142
and information exchange, 57, 143-145
and information needs, 138-139
and interaction, 141
and knowledge accumulation, 10
and language, 135
learning, 58
and literature, 38, 40-41, 139-140
and personal networks, 34
and time resources, 105
information translation, 48-49, 75, 85, 131
see also communication, translation
information work, ix, 29-30
innovation, 8, 99, 129
intellectual cultures, 70-72
styles of thought in, 76-77
interaction, 126-127
interdisciplinarity
definitions of, ix, 1, 20
and education, 20
and information needs, 32
and information organization, 138-142
management of, 131-132
risks of, 82-84

interdisciplinary institutions, 2-3, 64-66
ad hoc sites, 66
articulation work of, 100
and collaboration, 108, 111-113
communication in, 101, 112
and cross-disciplinary training, 84-85
and cross-disciplinary work, 64-65
and discipline-based structures, 98
and discipline building, 64-65, 113
goals of, 3-4, 65, 80-81
individual advancement in, 81-85
and intellectual cultures, 69-70
interaction in, 127
and leeway, 127-128
membership in, 3-5, 99, 115
research groups in, 25
research programs in, 65, 68
role of, 97
and shadow structures, 66-68
and technology, 113114
intermediaries, 47-50
and actor network theory, 47
graduate students as, 48-50
training of, 49-50
as translators, 48-49
interpersonal communication
see personal networks
Internet, 38, 125-126, 138, 144
invisible colleges, 37, 44, 57
see also invisible constituencies
invisible constituencies, 125-126

journals, 41, 43, 54-56, 137

knowledge, vii, 7-11, 141-142
and boundary crossing modes, 91-92
and communication, 8, 89-91
and interdisciplinary development, 86
and intermediaries, 141
interpretive, 86-87
levels of, 85-93
and overload, 93
as process, 9-10
strategies for development, 92-93
tacit, 86
units of, 7-8, 10-11

language, 12, 117-118
creolization of, 75-76
in cross-disciplinary learning, 108-110
jargon in, 134
mediation of, 134-135
and personal networks, 44
and research community, 6
vocabularies, 73-75
learning, 42-46, 129
leeway, 98-105, 127-128, 138
and alignment, 101
and articulation work, 99-100
defined, 98, 99
and export, 120
and information needs, 138-139
limits of, 101
symbolic, 103-104
technological, 104-105
libraries, 41, 135, 140, 142-143
literature, 37-42, 108-109
as accumulation tools, 137-138
and boundary crossing activity, 14-18
and boundary crossing modes, 39
citation analysis of, 119

electronic forms of, 40-41
emerging fields of, 80-81
experimental/theoretical, 107
footnote chasing in, 43
general vs. specific, 40, 44
language barriers in, 72-73
and learning, 43-44
methods of reading, 39-40
name recognition in, 43-44
organization of, 139-141
and overload, 94
placing, 54-55
standards of, 81-82
technical nature of, 53
unpublished, 42
see also audience, bibliometrics, core and scatter, journals, reading, review literature, writing

memberships, 25-26, 132-133
metaphors, 12, 143
multiple audiences, 51-52

online communication, 57
see also communication, electronic; e-mail
overload
see information overload

peer review, 80-82
personal networks, 32-35, 125-126
and communication, 33
and e-mail, 34
and overload, 94
probing, 31-32, 34, 42
problem-focused research, 19-20, 66, 98-99
productivity, 109-110
program groups, 14
concept of, 4
proximity, 101-103

reading, 39-42
research
climate, 114
credibility, 103-104
environments, 2-3, 6, 61
groupings, 4, 66-69, 138-141
levels, 12, 98-100
management, 4-5
organizations, 64-66
problems, 19, 98-99, 138
problem areas, 64, 66, 138-141
programs, 4-5, 25-26, 65-66, 68
rewards, 80-81, 114
trends, 79-80
work, 14, 21, 65, 123-125
see also work modes
research structures, 25-26, 64-71, 131-132
review literature, 41
as accumulation tools, 137-138
and import process, 41
and export process, 54

scatter, 15, 37-38
see also core and scatter
scholarly communication, 29, 35-42
scientific growth curve, 79-80
seeking
see information seeking
serendipity, 36, 40, 45, 111
shadow structures, 66-68
and interdisciplinarity, 131
social networks
see personal networks
social world, 98-101
specialization, 80-82
students, graduate, 6, 11-12, 49-50, 67, 83-85, 107-108, 114
as intermediaries, 48-50

and technology, 105
synthesis, 135-137

teaching, 46-47
team leaders, 21 fig, 22-23
information gathering, 31
technology, 104-105, 113-114
see also information technology
theoretical research, 52, 106-107
thought, styles of, 76-79, 117-118
time, 58, 101, 104-105
training, 20, 41, 49, 77, 83-85
and discipline structure, 84-85
interdisciplinary institutions, 83-85
trajection
see boundary trajection
transdisciplinarity, x, 1
translation, 14-15, 25, 47, 130, 133, 135
see also information translation

work location, 25-26, 114-115, 126-129
work modes, 18-24, 127-128
information organization, 140-141
see also knowledge modes, research work
writing, 50-56

MIX
Papier aus verantwortungsvollen Quellen
Paper from responsible sources
FSC® C105338

If you have any concerns about our products,
you can contact us on
ProductSafety@springernature.com

In case Publisher is established outside the EU,
the EU authorized representative is:
Springer Nature Customer Service Center GmbH
Europaplatz 3, 69115 Heidelberg, Germany

Printed by Libri Plureos GmbH
in Hamburg, Germany